New Monte Carlo Methods with Estimating Derivatives

New Monte Carlo Methods with Estimating Derivatives

G.A. Mikhailov

///VSP///
Utrecht, The Netherlands, 1995

VSP BV
P.O. Box 346
3700 AH Zeist
The Netherlands

First published in 1995

ISBN 90-6764-190-1

CIP-DATA KONINKLIJKE BIBLIOTHEEK, DEN HAAG

Mikhailov, G.A.

New Monte Carlo methods with estimating derivatives / G.A.
Mikhailov. - Utrecht : VSP
With ref.
ISBN 90-6764-190-1 bound
NUGI 811
Subject headings: Monte Carlo methods / mathematical
physics.

Printed in The Netherlands by Koninklijke Wöhrmann bv, Zutphen

CONTENTS

PREFACE

It is possible to use weighted Monte Carlo methods for solving many problems of mathematical physics (boundary value problems for elliptic equations, the Boltzmann equation, radiation transfer and diffusion equations). In this case an appropriate Markov chain is simulated on a computer and necessary functionals are estimated using the weight which, after each transition, is multiplied by the ratio of the integral equation kernel and the transition density function.

Weight estimates make it possible to evaluate special functionals, for example, derivatives with respect to parameters of a problem. The triangular system of integral equations is obtained by a multiple differentiation with respect to a parameter of the original integral equation.

In this book new weak conditions are presented under which the corresponding vector Monte Carlo estimates are unbiased and their variances are finite. For Helmholtz and radiation transfer equations, new recurrent scalar representations of these estimates are constructed.

Based on linearization, parametric derivatives can be used for solving reverse, stochastic and nonlinear problems.

It appears that by using direct differentiation of Helmholtz and transfer equations, it is possible to construct effective iterative processes for estimating critical parameters of these equations. Corresponding new algorithms with numerical tests are considered in detail. In addition, parametric derivatives for the Helmholtz equation represent solutions of metaharmonic equations.

Consequently, the author has constructed new Monte Carlo methods for solving the Helmholtz equation with a nonconstant parameter, including the stationary Schrodinger equation. The corresponding estimates are of the power type with respect to the parameter. This is especially suitable for differentiation.

Using Green functions, it is sometimes possible to make the weight dependent on a spatial point and to estimate corresponding derivatives. On the other hand, it is possible to construct similar estimates on the basis of a sufficiently regular extension of standard point estimations of the solution. Based on the imbedding theorem these estimates allow one to optimize Monte Carlo algorithms for a global estimate of the solution, for instance, in uniform metric. In this book corresponding new results for linear and nonlinear problems are presented. These results, in particular, demonstrate the advantage of the 'path estimate' as compared to the 'collision estimate' for solving linear and nonlinear radiation transfer equations in total.

In constructing these algorithms under study, the method of additional randomization is especially important. Therefore, some methods of random function simulation are considered in the special appendix. A new method of substantiating and optimizing the recurrent Monte Carlo estimates without using the Neumann series are presented in the Introduction. This method appears especially effective for substantiating Monte Carlo algorithms with branching, which is used when solving nonlinear problems.

This book will be of value and interest to specialists in the field of computational mathematics and physics, probability theory and mathematical statistics.

Chapter 1

Estimation of integrals and solution of integral equations

1.1. ESTIMATION OF INTEGRALS

Monte Carlo estimates for integrals are constructed on the basis of the relations (see, for example, [17,21,24]):

$$J = \int_Y g(y)dy = \mathrm{E}\xi, \quad \xi = g(\eta)/p(\eta),$$

where $p(y),\ y \in Y$, is the distribution density for the random vector η, and

$$\mathrm{D}\xi = \mathrm{E}\xi^2 - J^2 = \int_Y \frac{g^2(y)dy}{p(y)} - J^2.$$

To estimate the expected value of ξ, one usually uses a sequence $x_1, ..., x_N$ of independent samples of ξ obtained by Monte Carlo. Then, by the law of large numbers,

$$\mathrm{E}\xi \approx \tilde{x}_N = \frac{1}{N}\sum_{k=1}^{N} x_k, \tag{1.1}$$

$$\mathrm{D}\xi = \sigma^2 \approx \tilde{\sigma}^2 = \frac{N}{N-1}\left(\frac{1}{N}\sum_{k=1}^{N} x_k^2 - \tilde{x}_N^2\right).$$

To evaluate the error of estimate of Eq. (1.1), a confidence interval is usually calculated which contains $\mathrm{E}\xi$ with probability $1-\varepsilon$, where $0 < \varepsilon \ll 1$. The central-limit theorem can be used to construct the confidence interval approximately, provided that $\mathrm{D}\xi < +\infty$. For N sufficiently large, the following asymptotic equality holds

$$P(|\tilde{x}_N - J| < \beta\sigma/N^{1/2}) = P\left(\frac{\sqrt{N}}{\sigma}|\tilde{x}_N - J| < \beta\right) \approx \frac{1}{\sqrt{2\pi}}\int_{-\beta}^{\beta} e^{-(x^2/2)}dx = p(\beta).$$

Thus, setting $p(\beta) = 1 - \varepsilon$, we obtain the desired confidence interval: $\tilde{x}_N \pm \beta\sigma/(N^{1/2})$. In practice, σ is usually unknown as well, and sample estimates $\tilde{\sigma}$ must be used for it; this is well, if N is sufficiently large. The actual values of β are taken to be 1, 2 or 3, which correspond to the confidence levels $p(\beta) = 0.68,\ 0.95,\ 0.997$, respectively. As a rule, the same trajectories are used to estimate a number of functionals for which the sample estimates are obtained as total contributions from various collisions. Clearly, it is not efficient in such calculations, to estimate $\tilde{\sigma}$ by summation of x^2 over all trajectories. Therefore, the trajectories are divided up into a few groups each of which has M trajectories; from those data, the standard deviation of the random variable is estimated

$$\zeta_M = \frac{1}{M}\sum_{i=1}^{M} \xi_i \,,$$

where ξ_i are independent samplings of ξ. Let $\sigma_{M,S}$ be a statistical estimate of the quantity $\sigma_M = (\mathrm{D}\zeta_M)^{1/2}$ calculated for S groups ($N = MS$). Then the standard error of the quantity $\tilde{x}_N$ can be estimated

$$\sigma(\tilde{x}_N) = \frac{\sigma}{\sqrt{N}} \approx \frac{\tilde{\sigma}_{M,S}}{\sqrt{S}} \,.$$

Sometimes it is necessary to use, in the Monte Carlo method, estimates that have infinite variance (see [39]). Let $F_\xi(y)$ be a distribution function of the random variable ξ and let

$$U(x) = \int_0^x y^2 dF_\xi(y).$$

If the variance $\mathrm{D}\xi$ diverges as $\ln x$ (i.e., $U(x) \sim c \cdot \ln x$), then the random variable

$$\gamma_M = \left[\frac{M}{\ln M}\right]^{1/2} (\zeta_M - \mathrm{E}\xi),$$

is asymptotically normal $(0, d)$, where d is a standard deviation [19]. The standard error of the resultant estimate is given by

$$d^*(\zeta_{MS}) = d^* \left[\frac{\ln(MS)}{M \cdot S}\right]^{1/2} ,$$

which differs from the analogous quantity for the case of bounded variance by the factor $[\ln(MS)/\ln M]^{1/2}$. Although the variance of ζ_{MS} is infinite the foregoing limit theorem enables us to construct approximately a confidence interval for $\mathrm{E}\xi$, as was explained for the normal distribution with $d = d(\zeta_{MS})$. Indeed, for large $N = M \cdot S$, the divergence of $D\zeta_M$ is due to the tail of distribution which does not essentially affect the observation. This supposition agrees with the results of calculations. Therefore, the rate of probability convergence of estimate ζ_N is given by $[\ln N/(N)]^{1/2}$.

The Monte Carlo method is believed to be rather efficient for estimation of a small number of linear functionals of the solution which are expressed by integrals, in general, of infinite multiplicity (the effect of arguments usually exponentially decreases, this providing a possibility of constructing a sufficiently efficient statistic estimate). Using as an example the estimates $\{\xi\}$ of a functional J we will show the main approach to minimizing the cost, i.e. the computer time or an average number of operations which is necessary for obtaining a given error δ. We assume that the statistical estimate error is measured by the standard deviation as when the number of samples increases the confidence intervals used in practice are, as a rule, asymptotically proportional to this quantity. We consider first unbiased estimates, i.e. we assume that $\mathrm{E}\xi = J$.

Thus, let a family of random variables $\{\xi\}$ with finite variances $\mathrm{D}\xi$ be given which are such that $\mathrm{E}\xi = J$. Using n independent samples we construct random estimates of the form

$$\zeta_n = \frac{\sum_{i=1}^{n} \xi^{(i)}}{n}$$

such that

$$\mathrm{E}\zeta_n = J, \quad \mathrm{D}\zeta_n = \frac{\mathrm{D}\xi}{n} .$$

Denote by t an average cost of realization of random variable ξ, i.e. an average cost of one random sample. Then the minimization of the cost of the estimates considered is equivalent to the following minimization problem:

$$S = nt \to \min, \quad \frac{\mathrm{D}\xi}{n} = \delta^2$$

or $t\mathrm{D}\xi \to \min$. The cost minimization in the class of unbiased estimates with a finite variance is thus independent of the required error δ and is also invariant to n. Note that $t\mathrm{D}\xi$ as a quality criterion for the unbiased estimate of the Monte Carlo method has been introduced in [23].

Consider now a monoparametric family of biased estimates ξ_ε with the cost $t(\varepsilon)$, variance $d(\varepsilon)$ and bias of order ε, that is

$$|\mathrm{E}\zeta_\varepsilon - J|^2 \lesssim c\varepsilon^2.$$

Symbol $\lesssim$ means that an upper estimate of bias of order ε is used in the general case. The root mean square of deviation is defined here by the relation

$$\mathrm{E}(\xi_\varepsilon - J)^2 = \mathrm{D}\xi_\varepsilon + (J - \mathrm{E}\xi_\varepsilon)^2.$$

Hence,

$$\mathrm{E}(\zeta_{\varepsilon,n} - J)^2 \lesssim \frac{d(\varepsilon)}{n} + c\varepsilon^2. \tag{1.2}$$

The cost minimization is thus equivalent here to the following constrained minimization problem:

$$S_\varepsilon = nt(\varepsilon) \to \min_{N,\varepsilon}, \quad \frac{d(\varepsilon)}{n} + c\varepsilon^2 = \delta^2$$

or

$$\frac{t(\varepsilon)d(\varepsilon)}{\delta^2 - c\varepsilon^2} \to \min_{\varepsilon}, \quad c\varepsilon^2 < \delta^2.$$

By solving this problem we obtain optimal values n and ε as the functions of δ which determines a certain relation between n and ε. The equalization of the terms on the right-hand side of expression of Eq. (1.2), which is sometimes used in practice, is thus sure to yield no optimal variant of the estimate. A simple analysis however shows that the equalization technique yields an optimal order of the cost if $t(\varepsilon) = \text{const}$ and $d(\varepsilon) \to \infty$ for $\varepsilon \to 0$.

In order to construct a global estimate of the function

$$\varphi(x) = \int_Y g(x,y)P(dy)$$

in the bounded domain D, one can estimate it's values in nodes of a rectangular grid with step H by the Monte Carlo method and then perform linear filling. We denote the estimate thus obtained by $\tilde{\varphi}(x)$. Generally speaking, the function $\tilde{\varphi}(x)$ is a random field, whose distribution is due to the sample size (i.e. the number of realizations of the Monte Carlo estimates), to the way of constructing the estimate ξ, and to step h.

The problem on the convergence of $\tilde{\varphi}$ to φ in some metric can be solved by considering the quantity

$$B(\varphi,\tilde{\varphi}) = \mathrm{E}\|\varphi(x) - \tilde{\varphi}(x)\|_{L(D)}\,,$$

where $L(D)$ is a corresponding Banach space (see, for example [3]). This problem is connected with the complexity of the estimation algorithm, i.e. with the average number of operations that provides the validity of the inequality $B(\varphi,\tilde{\varphi}) < \delta$. Estimating $B(\varphi,\tilde{\varphi})$ is the simplest for the space $L_2(D)$:

$$\begin{aligned} B^2(\varphi,\tilde{\varphi}) &= \left(\mathrm{E}\left\{\int_D [\varphi(x) - \tilde{\varphi}(x)]^2 dx\right\}^{1/2}\right)^2 \\ &\le \int_D \mathrm{E}[\varphi(x) - \tilde{\varphi}(x)]^2 dx = \int_D \mathrm{D}\tilde{\varphi}(x)dx + \int_D [\varphi(x) - \tilde{\varphi}(x)]^2 dx. \end{aligned} \tag{1.3}$$

In particular, expression (1.3) confirms the importance of the uniform minimization of the value $\mathrm{D}\xi$. We assume that the second-order derivatives of the function $\varphi(x)$ are uniformly bounded in D. Then

$$B^2(\varphi,\tilde{\varphi}) \le d/n + C_0 h^4.$$

Therefore the problem of minimizing the complexity can be here formulated in the form:

$$S_0 = nt_0 h^{-k} \to \min_{n,h}, \quad d/n + c_0 h^4 = \delta^2, \tag{1.4}$$

where k is the phase space dimension, n is the sample size, h is the grid step, the meaning of d is seen from (1.3), and $t_0 = t\,\mathrm{mes}(D)$, where t is the input for one realization of ξ. Problems of the form of (1.4) are considered in detail in [38,39]. The optimal order of values h, n and S_0 is as follows:

$$h_0^* \asymp \delta^{1/2}, \quad n_0^* \asymp \delta^{-2}, \quad S_0^* \asymp \delta^{-(2+k/2)}. \tag{1.5}$$

In this case the estimate $B(\varphi,\tilde{\varphi}) < \delta$ is valid and therefore $\tilde{\varphi}$ converges to φ in the metric L_2.

One can obtain similar results for the metric $C(D)$ by using the theorems of embedding of the space $W_2^l(D)$ in the space $C(D)$, provided $2l > k$. This condition implies that using the first-order derivatives suggests $k = 1$, i.e. allows one to consider the convergence of the estimate of the solution of a one-dimensional equation (or a multi-dimensional equation in a given straight line). In this case the inequality holds

$$\|\varphi - \tilde{\varphi}\|_C^2 < K\Big[\int_D [\varphi(x) - \tilde{\varphi}(x)]^2 dx + \int_D [\varphi'(x) - \tilde{\varphi}'(x)]^2 dx\Big].$$

Since the variance of the difference of independent random quantities is equal to the sum of their variances, one must use the following inequality when making an independent estimate of values at nodes of the given grid:

$$\mathrm{D}\tilde{\varphi}'(x) < \frac{d(x)}{nh^2}\,.$$

In addition, since $\tilde{\varphi}'$ is a step function, one needs to consider the relation

$$\int_D [\varphi'(x) - \tilde{\varphi}'(x)]^2 dx < C_1' h^2.$$

Thus, it is appropriate to consider the following estimate asymptotically for $h \to 0$:

$$\|\varphi - \tilde{\varphi}\|_C^2 < \frac{d_1}{nh^2} + C_1 h^2.$$

This estimate leads to the following problem of complexity minimization:

$$S_1 = nt_0 h^{-k} \to \min_{n,h}, \quad \frac{d_1}{nh^2} + C_1 h^2 = \delta^2. \tag{1.6}$$

On substitution of $n' = nh^{-1}$ problem (1.6) reduces to the problem on uniform pointwise minimization of complexity of the histogram method for $k = 3$, discussed in [38,39]. The optimal values are of the following order of magnitude:

$$h_1^* \asymp \delta, \quad n_1^* \asymp \delta^{-4}, \quad S_1^* \asymp \delta^{-5}.$$

Thus, in a one-dimensional case the complexity of global estimation of the solution in $C(D)$ is quadratic with respect to the estimate in $L_2(D)$. The complexity of the estimation in $C(D)$ can be considerably reduced by using a dependent estimate of values of φ that provides the relation

$$|\tilde{\varphi}(x) - \tilde{\varphi}(x+h)| < Ch, \quad h \to 0$$

with the probability 1 [20]. In this case, instead of (1.6), we obtain the problem

$$S_1 = nt_0 h^{-1} \to \min_{n,h}, \quad d_2/n + C_2 h^2 = \delta^2, \tag{1.7}$$

where

$$h_2^* \asymp \delta, \quad n_2^* \asymp \delta^{-2}, \quad S_2^* \asymp \delta^{-3}.$$

There are various ways of correlating the estimates in the Monte Carlo method (see [17,20,39,40]).

In this connection a method proposed for solving the Laplace equations in [12] and for solving a general integral equation in [40] may be of special interest. The method involves the recursion

$$\xi_x = h(x) + \frac{k(x, x_1)}{p(x, x_1)} \xi_{x_1}$$

that is a standard estimate ξ, provided that $x_0 = x$ [40].

Let us assume that the first step $x \to x_1$ is realized with respect to the density $p_0(x_1)$, which is independent of x and such that $p_0(x_1) \neq 0$ if $k(x, x_1) \neq 0$ at least for some $x \in D$.

Then the random quantity

$$\xi_x^{(0)} = h(x) + \frac{k(x, x_1)}{p_0(x_1)} \xi_{x_1}$$

is an unbiased estimate of the solution $\varphi(x)$, i.e.

$$\mathrm{E}\xi_x^{(0)} = \varphi(x) \quad \forall x \in D .$$

The following formula is valid [40]:

$$\mathrm{D}\xi_x^{(0)} = \int_D \frac{k^2(x, x_1)}{p(x_1)} \mathrm{E}\xi_{x_1}^2 \, dx_1 - [\varphi(x) - h(x)]^2$$

that also shows the importance of uniform minimization of $\mathrm{E}\xi^2$. On the basis of this formula a minimax variant of the density $p_0(x)$ was constructed in [40]. It is obvious that the values $\xi_x^{(0)}$ are strongly dependent for different x, since the value of ξ_{x_1} is the same for them. If the function

$$k_0(x) = k(x, x_1) \quad \forall x_1 \in D$$

is sufficiently smooth, one can prove the convergence of the corresponding estimate $\tilde{\varphi}(x)$ in the metric $C(D)$ by using the theorems of embedding from $W_2^l(D)$ in $C(D)$ with $l > 1$, if a corresponding smooth filling $\tilde{\varphi}(x)$ is constructed. For example, in order to use $W_2^2(D)$ one must construct a proper quadratic filling; in this case for $k < 4$ the optimal variant $\tilde{\varphi}$ is determined by formulae (1.7) with substitution of h^{-1} for h^{-k} in the expression for S_1, and

$$h_3^* \asymp \delta, \quad n_3^* \asymp \delta^{-2}, \quad S_3^* \asymp \delta^{-(k+2)}.$$

In addition, here one needs the uniform boundedness of the third-order derivatives of the solution. Simple considerations show, that if a smooth quadratic estimate $\tilde{\varphi}$ is constructed by using independent values at the grid nodes, then the complexity is minimized by using formulae of form of (1.6) with substitution of h^{-k} for h^{-1} in the expression for S_1 and substitution of $d_1/(nh^2)$ for $d_3/(nh^2)$ in the expression for δ^2. This leads to the following asymptotic expressions:

$$h_4^* \asymp \delta, \quad n_4^* \asymp \delta^{-6}, \quad S_4^* \asymp \delta^{-(k+6)},$$

where $k \leq 3$.

The standard use of pseudorandom numbers 'in succession' gives no universal possibility for constructing correlated samples, because of the possible differences in the length of the trajectories in different calculations. However, it is possible to divide beforehand the sequence $\{\alpha_n\}$ into large sub-sequences and use them to simulate the corresponding trajectories. This procedure is simple when the multiplicative congruental method [24]

$$u_0 = 1, \quad u_n = u_{n-1}M(\mathrm{mod}\, 2^m), \quad \alpha_n = u_n 2^{-m}$$

is used. In this method

$$\alpha'_n = \alpha_{kn} = \{(M^k)^n 2^{-m}\},$$

i.e., the initial number of the parts of length k are also constructed by the multiplicative generator with the factor M^k, which is defined by the value of α_k (since only m lowest digits are used). Thus, the simulation of the trajectories in different variants begins with the same pseudonumbers which lead to the coincidence (or similarity) of significant parts of the trajectories

On the basis of the Kolmogorov criterion in [20] the convergence of estimates in the dependence sampling technique was investigated; in this method the values of a function (e.g., an integral with a parameter) are calculated using the same set of samplings.

Let $f(x) = \mathrm{E}\xi(x,\omega)$ and $\varepsilon(x,\omega) = \xi(x,\omega) - f(x)$. Say that $\xi \in Gamma_k$ if

$$\left|\frac{\partial^k \varepsilon(x,\omega)}{\partial x^k}\right| \le G_k(\omega),$$

$$\int [G_k(\omega)]^p dP(\omega) = c_k < \infty, \quad p \ge 2.$$

Theorem 1.1. *If the function $\xi(x,\omega)$ belongs to all $Gamma_k (k = 1, ..., l)$, then the random vector function*

$$\varphi_M = \sqrt{M}\left\{M^{-1}\sum_{i=1}^{M}\frac{\partial^k \xi(x,\omega_i)}{\partial x^k} - f^{(k)}(x)\right\}, \quad k = 0, ..., l-1$$

weakly (see appendix) converges to a continuous Gaussian vector function with zero mean and the covariance matrix

$$B_{jk}(x',x') = \mathrm{M}\{[\xi^{(j)}(x',\omega) - f^{(j)}(x')][\xi^{(k)}(x',\omega) - f^{(k)}(x')]\}$$

(*see* [20]). □

1.2. RECURRENT ESTIMATES OF MONTE CARLO METHOD FOR THE SOLUTION TO AN INTEGRAL EQUATION OF THE SECOND KIND

Let's consider the integral equation

$$\varphi(x) = \int_X k(x,x')\varphi(x')dx' + h(x) \tag{1.8}$$

or in the operator form

$$\varphi = K\varphi + h,$$

where $X \in R_k$ is compact set, $h \in L_\infty(X)$, $K \in [L_\infty \to L_\infty]$. Here L_∞ is the space with the norm

$$\|h\| = \mathrm{vrai}\sup_x \|h(x)\| < \infty,$$

and

$$\|K\| = \operatorname{vrai}\sup_x \int_X |k(x,x')|dx',$$

under the condition that the kernel $k(x, x')$ is a measurable function. This relation holds e.g., if $k(x,x') \geq 0$. In general case

$$\|K\| \leq \operatorname{vrai}\sup_x \int_X |k(x,x')|dx',$$

For simplicity we shall omit the symbol vrai and the expression 'almost everywhere with respect to Lebesgue measure' if there will not arise a misunderstandness of utilization of the space L_∞. The kernel $k(x, y)$ can be quite arbitrary, but it is necessary to satisfy the condition that $K \in [L_\infty \to L_\infty]$.

Note that generalized factors like 'delta functions' are allowed in the kernel $k(x, x')$; in so doing the same factors are necessary in the transition density of the Markov chain discussed below (also see [17]). These generalized factors are not considered in the corresponding random weight, i.e. weight factors are constructed as ratios of measurable factors of the kernel to the corresponding factors of the transition density.

If $k(x, y) \geq 0$, then it is convenient sometimes to consider the equation (1.8) in the cone of nonnegative functions. In Section 2.1 an equation for the covariance matrix of the weighted vector estimate will be considered on a more complicated cone of functions whose values lie in the space of positive definite matrices. The corresponding operators will be positive. These operators will be used only in Chap. 2, therefore we shall give there some information needed. Here we just note that a cone is defined as a convex set which includes the zero element and possesses following property: if non-zero x belongs to the cone then $-x$ does not belong to the cone.

If the kernel is regular (the corresponding conditions are well known, see, e.g., [10]), the integral operator K is compact.

Spectral radius of the operator K is often used in investigating the accuracy and the computational cost of the weighted Monte Carlo methods and is defined as follows [10]:

$$\rho(K) = \lim_{n\to\infty} \|K^n\|^{1/n},$$

and

$$\rho(K) \leq \|K^n\|^{1/n}, \quad n = 1, 2, \dots .$$

If the operator K is compact and in the invariant subspaces corresponding to eigen-values whose modula are equal to $\rho(K)$ there are no addition elements, then following inequality holds [10]:

$$\|K^n\| \leq c[\rho(K)]^n, \quad n = 1, 2, \dots .$$

In general case [10]

$$\|K\| \leq c(\varepsilon)[\rho(K) + \varepsilon]^n \quad \forall \varepsilon > 0.$$

If $\rho(K) < 1$, then the solution to (1.8) can be represented in the form of a convergent Neumann series:

$$\varphi = \sum_{n=0}^{\infty} K^n h .$$

Simple estimations of the solution $\varphi(x)$ to (1.8) with nonnegative entries can be obtained on the basis of this relation [40, Lemma 1.1]: if

$$k(x,x') \geq 0, \quad h(x) \geq 0, \quad \psi(x) \geq 0, \quad \rho(K) < 1,$$

then from $\psi \leq K\psi + h$, $\psi \geq K\psi + h$ inequalities $\varphi \geq \psi$, $\varphi \leq \psi$ follow.

As the basic case we here consider nonnegative $k(x,x')$, $h(x)$ and $\varphi(x)$, since going to the case with alternating signs is performed by a standard method based on an appropriate majorization (see [17,40] and further discussion). We will show below that all the results obtained can be used for constructing estimates of the functionals (f, φ), where $f \in L_1(X)$, i.e. $\int |f(x)|dx < +\infty$.

In this section we assume that the spectral radius $\rho(K) < 1$. In so doing the solution to equation (1.8) is expressed as the Neumann series, but here we construct and investigate the basic recurrent estimate of the Monte Carlo methods by means of a direct probabilistic approximation of equation (1.8). We show in the following sections that this approach is especially effective with branching Markov chains, where the complete expression of the estimate is of necessity cumbersome (see, for instance, [17]). In addition, the recurrent representations of estimates and second-order moments allow one to use the Bellman principle for their optimization. In so doing the corresponding known results of the theory of such estimates [17,40] are clarified, refined, and generalized. In the present section the sufficiency of the application of the Bellman principle is additionally justified.

Here the appropriate parametrization of the initial problem is of great importance. For equation (1.8) that means passing to the equation

$$\varphi_\lambda = \lambda K \varphi_\lambda + h, \quad 0 < \lambda < 1/\rho(K). \tag{1.9}$$

Clearly, the solution to the equation $\varphi_\lambda \equiv \varphi(x,\lambda)$ depends on λ monotonously and even strictly monotonously, except for trivial degenerated cases.

To construct the estimate of the Monte Carlo method we introduce the terminating Markov chain $\{x_n\}$, $n = 0, 1, \dots N$, with the transition density $p(x,x')$, where the quantity

$$p(x) = 1 - \int_X p(x,x')dx' \geq 0$$

is considered as the probability of the trajectory terminating in the transition from x to x'; N is a random number of the last state of the chain. The Markov chain is related to equation (1.8) by the weight factor $q(x,x')$. For the transition from x to x' this factor is calculated by the formula

$$q(x,x') = \frac{k(x,x')}{p(x,x')} \tag{1.10}$$

and $q(x,x') = 0$ if this transition terminates the trajectory. To simplify the mathematics we can formally assume that after termination trajectory gets into a special absorbing point and remains there. Relation (1.10) implies, in particular, that the following known condition [17,40] should be met:

$$p(x,x') \neq 0 \quad \text{for} \quad k(x,x') \neq 0. \tag{1.11}$$

We have to construct an unbiased random estimate of the solution for the chain originating at the point x, i.e. to construct such a functional ξ that $\mathrm{E}\xi = \varphi(x)$. Obviously, this estimate allows the following randomization of equation (1.8):

$$\xi_0 = h_0 + q(x_0, x_1)\xi_1, \tag{1.12}$$

where $h_0 = h(x_0)$ and $\xi_i = \xi$ for $x = x_i$; one can easily see that

$$\begin{aligned} \mathrm{E}[q(x_0, x_1)\xi_1] &= \mathrm{EE}[q(x_0, x_1)\xi_1|x_1] \\ &= \mathrm{E}[q(x_0, x_1)\psi(x_1)] = [K\psi](x_0), \quad \psi \equiv \mathrm{E}\xi, \end{aligned}$$

i.e. relation (1.12) is a probabilistic approximation of equation (1.8) in the sense that in average (1.12) coincides with (1.8). However, recurrent estimate (1.12) is unbiased only if the additional condition is met:

$$\mathrm{E}\xi < C < +\infty \quad \forall x \in X \tag{1.13}$$

because the solution to equation (1.8) is unique in L_∞. It is interesting to note that here we have an assertion similar to the known Lax theorem: from the probabilistic approximation and stability of (1.13) follows the convergence (in the sense of validity of the law of large numbers). In order to deduce the unbiasedness of ξ directly from (1.12), we note that when continuing recursion (1.12) up to the end of the trajectory $\{x_n\}$ we obtain the known representation of the 'collision' estimate ξ [17,40]. For parametrized equation (1.9) the representation is of the form:

$$\xi_0(\lambda) = \sum_{n=0}^{\infty} \lambda^n Q_n h(x_n), \tag{1.14}$$

where

$$Q_0 \equiv 1, \quad Q_n = Q_{n-1} q(x_{n-1}, x_n).$$

We will point out rather apparent properties of the function $\mathrm{E}\xi(\lambda)$. As a function of λ (at any fixed $x = x_0$), it is continuous up to $\psi = \infty$, since this function is the expectation, i.e. integral of monotonous with respect to λ and nonnegative function of trajectory over probabilistic measure. The integral $\mathrm{E}\xi(\lambda)$ allows differentiation under the sign E, because as a result one obtains an integral of nonnegative monotone function. Therefore $\mathrm{E}\xi(\lambda)$ is a convex function of λ and (see, for example, [46]) $\|\xi(\lambda)\|$ is a convex function of λ. Further, since X is a compact set, the same property is valid for the supremum of $\|\xi(\lambda)\|$ from the continuity of $\mathrm{E}\xi(\lambda)$ at $+\infty$ for any x. That means that there exists such a sequence $\{\lambda_k\}$, that

$$\lim_{k\to\infty} \|\xi(\lambda_k)\| = +\infty, \quad \|\xi(\lambda_k)\| < +\infty .$$

Therefore there exist such values λ_k that the value of $\|\xi(\lambda_k)\|$ is as large as one likes but finite, and $\|\xi(\lambda)\|$ is continuous if $\lambda < \lambda_k$ [46] as it is a convex function.

The experience of solving various problems by means of the Monte Carlo method shows that one can always construct such a function $p(x, x')$ that the following useful condition is met:

$$|q(x, x') < C < +\infty . \tag{1.15}$$

Theorem 1.2. *If $\rho(K) < 1$, $k(x, x') \geq 0$, $h(x) \geq 0$ and conditions (1.11), (1.15) are met, then recursion (1.12) defines the unbiased estimate ξ.*

Proof. The function $\|E\xi(\lambda)\|$ is monotonous and continuous with respect to λ up to $+\infty$ and by virtue of condition (1.15) we have $\|E\xi(\lambda)\| < +\infty$ for sufficiently small $\lambda > 0$. Therefore if $\|E\xi\| = +\infty$, then one can find such $\lambda < 1$ that $\|\varphi\| < \|E\xi(\lambda)\| < +\infty$.

However, in so doing we have $E\xi(\lambda) = \varphi_\lambda$, but $\|\varphi_\lambda\| \leq \|\varphi\|$ due to the monotonous dependence of φ_λ on λ. The controversy obtained proves Theorem 1.2. □

Theorem 1.2 is a well-known statement [17,40], but it is of interest because in its proof one uses only the recurrent nature of the estimate and does not perform cumbersome calculations that involve detailing the Neumann series and preliminary averaging the complete expression for ξ over 'absorbing-surviving' [17].

Theorem 1.2 can be obviously extended to the case with alternating signs, provided the condition $\rho(K_1) < 1$ is met, where K_1 is an integral operator with the kernel $|k(x, x')|$, since in this case we have

$$|\xi_0| \leq \xi_0^{(1)} = |h_0| + |q(x_0, x_1)|\xi_1^{(1)}$$

which means that $|E\xi| < C < +\infty$ and direct averaging of relation (1.12) gives the unbiasedness of ξ.

1.3. VARIANCE OF THE BASIC UNBIASED ESTIMATE

The variance $D\xi = E\xi^2 - \varphi^2$ is determined by the value $E\xi^2$ that depends significantly on the choice of the transition density $p(x, x')$. Let us consider the relation

$$\xi_0^2 = h_0^2 + 2h_0 q(x_0, x_1)\xi_1 + q^2(x_0, x_1)\xi_1^2 \,. \tag{1.16}$$

We can obtain from (1.15) a simpler recursion if we replace the term $2h_0 q(x_0, x_1)\xi_1$ by the corresponding expectation

$$E[2h_0 q(x_0, x_1)\xi_1] = 2h_0[K\varphi]_0 = 2h_0(\varphi_0 - h_0).$$

Let us show that such sibstitution does not affect the value of the mean square. Considering (1.16) as a recursion for ξ^2 and continuing it to the end of a trajectory, we obtain the expression

$$\xi_0^2 = \sum_{n=0}^{\infty} Q_n^2[h_n^2 + 2h_n q(x_n, x_{n+1})\xi_{n+1}].$$

When averaging this expression one can (see [17,40]) replace the bracketed term by its conditional expectation at a fixed value of x_n, i.e. by the quantity $h_n(2\varphi_n - h_n)$.

Thus, we have $E\xi^2 = E\zeta^2$ and

$$\zeta_0^2 = h_0(2\varphi_0 - h_0) + q^2(x_0, x_1)\zeta_1^2 \,. \tag{1.17}$$

Therefore, if the conditions of Theorem 1.2 are satisfied and $p(K_p) < 1$ (K_p is an integral operator with the kernel $k^2(x, x')/p(x, x')$), then the function $\psi \equiv E\xi^2$ is determined by the equation

$$\psi = h(2\varphi - h) + K_p\psi \tag{1.18}$$

which was first obtained in [9] when constructing Monte Carlo methods for solving systems of algebraic equations. One can easily show (see [40]) that $\|K_1\| < \|K_p\|$, hence under the mentioned conditions (1.18) is also valid in the case with alternating signs. In the case of constant signs, the most general result is as follows: the Neumann series for equation (1.18) determines the quantity $E\xi^2$ for the unbiased estimate ξ whether or not the condition $p(K_p) < 1$ [40] is met.

Further we will show the way to optimize the choice of $p(x, x')$ on the basis of the recurrent formulae for $E\xi^2$ and the Bellman principle. To do this, let us consider more general problem of minimization of the value of average variance when solving many equations with different $h^{(t)} \equiv h(x,t)$, $t = 1, ..., m$, namely minimization of the quantity

$$D(\Lambda) = \sum_{t=1}^{m} \lambda_t D\xi^2(t) = \sum_{t=1}^{m} \lambda_t E\xi^2(t) - \sum_{t=1}^{m} \lambda_t \varphi^{(t)2}.$$

Here $\varphi^{(t)} = K\varphi^{(t)} + h^{(t)}$, $\xi(t)$ is the corresponding unbiased estimate,

$$0 \le \lambda_t \le 1, \quad \sum_{t=1}^{m} = 1, \quad \Lambda = (\lambda_1, ..., \lambda_m).$$

Thus, the problem of minimization of $D(\lambda)$ reduces to minimization of the quantity

$$\sum_{t=1}^{m} \lambda_t E\xi^2(t) = E\zeta^2(\Lambda), \quad \zeta^2(\Lambda) = \sum_{t=1}^{m} \lambda_t \xi^2(t)$$

in this case

$$E\zeta_0^2(\Lambda) = \sum_{t=1}^{m} \lambda_t h_0^{(t)}(2\varphi_0^{(t)} - h_0^{(t)}) + K_p[E\zeta_1(\Lambda)] = a_0(\Lambda) + K_p[E\zeta_1(\Lambda)].$$

Continuing this recursion, we obtain for $n \ge 0$:

$$E\zeta_0^2(\Lambda) = \sum_{k=1}^{m} K_p^k[a_{n-k}(\Lambda)] + K_p[E\zeta_{n+1}^2(\Lambda)]. \tag{1.19}$$

In order to make use of the Bellman principle we will seek for the optimal chain in the class of inhomogeneous Markov chains. We assume the transitional density $p(x_n, x_{n+1}) = p_n(x_n, x_{n+1})$ to depend on n. In this case

$$E\xi_0^2(\Lambda) = A_n(\Lambda) + \int_X \frac{k^2(x,x')}{p_n(x,x')}\psi_{n+1}(x')dx', \tag{1.20}$$

where $A_n(\Lambda)$ is independent of p_n, and $\psi_{n+1} \equiv E\zeta_{n+1}^2(\Lambda)$. By analogy with the Bellman principle one can conclude that in the optimal case the function $p_n(x, x')$ must minimize the integral from (1.20). Thus, taking into account the principle of 'importance sampling' [17,40], the following relation must be valid:

$$p_n(x,x') = \frac{|k(x,x')|\psi_{n+1}^{1/2}(x')}{\int |k(x,x')|\psi_{n+1}^{1/2}(x')dx'}$$

and therefore

$$K_p[E\zeta_{n+1}^2(\Lambda)] = \left[\int |k(x,x')|\psi_{n+1}^{1/2}(x')dx'\right]^2.$$

Thus, the following infinite system of equations must be valid for the optimal case:

$$\psi_n = a_n(\Lambda) + (K_1\psi_{n+1}^{1/2})^2 \tag{1.21}$$

or $\psi_n^{1/2} = G_n(\psi_{n+1}^{1/2})$, where

$$G_n(\psi_{n+1}) = [a_n(\Lambda) + (K_1\psi_{n+1}^{1/2})^2]^{1/2}.$$

Let us assume that $\|K\| < 1$ and $a_n(\Lambda) \geq 0$.

Then the operator G_n is a contraction operator in L_∞ with the contraction factor $q \leq \|K\|$ (see, for example [40]). Therefore the solution of system (1.21) is existent and unique in $\mathbf{L}_\infty$, with $\psi_n \equiv \psi$ and

$$p_n(x,x') \equiv p(x,x') = \frac{|k(x,x')|\psi^{1/2}(x')}{[K_1\psi^{1/2}](x)}\ , \tag{1.22}$$

where ψ is the solution to the equation

$$\psi = \sum_{t=1}^{m} \lambda_t h^{(t)}(2\varphi^{(t)} - h^{(t)}) + (K_1\psi^{1/2})^2. \tag{1.23}$$

Note that the alternation of signs of the terms in the equation is not improbable here provided that the free term in (1.23) is nonnegative (this condition ensures the contraction property of the operator G).

To show the sufficiency of relations (1.22), (1.23) we must confirm that the equality $E\zeta^2(\Lambda) \equiv \psi$ is valid when using the transition density (1.22). It can be easily verified that the following relation is true in this case:

$$\psi = a + K_p\psi\ . \tag{1.24}$$

Nevertheless, this is not yet the asertion we are seeking for. Though the function $E\zeta^2(\Lambda)$ satisfies equation (1.24). It is not clear whether the solution of this equation is unique. It is known [40] that the function $E\zeta^2(\Lambda)$ is represented by the Neumann series for (1.24). On iterating (1.24) beginning with $\psi_0 \equiv \psi$, we arrive the expression

$$\psi \equiv \sum_{k=0}^{n} K_p^n a + K_p^{n+1}\psi\ .$$

Then on the basis of the inequality $K_1\psi^{1/2} \leq \psi^{1/2}$ that follows from (1.22), we obtain

$$K_p^{n+1}\psi \le \|K_1\|^{n+1} \to 0,$$

i.e. $\psi \equiv \mathrm{E}\zeta^2(\Lambda)$ is true if relations (1.22), (1.23) hold, which thereby define the unique optimal variant of the Markov chain with the proviso that $a \ge 0$.

For $m = 1$ this assertion was justified in an arbitrary case ($a(x) \not\ge 0$) in [5]. It is interesting to note that, as shown in [40], the unique solution to equation (1.23) coincides with φ in the case of constant signs. Hence one shows that connection between the 'value sampling' [17,40] used for optimizing the 'collision estimate' ξ and the standard 'importance sampling' [17,40] together with the Bellman principle.

Let us now assume that $x = (x^{(1)}, x^{(2)})$, then

$$\begin{aligned} k(x,x') &= k_1(x;x^{(1)'})k_2(x;x^{(1)'},x^{(2)'}) \\ p(x,x') &= p_1(x;x^{(1)'})p_2(x;x^{(1)'},x^{(2)'}). \end{aligned} \tag{1.25}$$

We consider the problem of optimizing only partial transition density p_2 (at a given density p_1) assuming that the optimal variant has the uniform Markov property. Taking into account representations (1.25) one can easily see that the following equality is valid:

$$[K_p\psi](x) = \int_{X_1} \frac{k_1^2(x;x^{(1)'})}{p_1(x;x^{(1)'})}\left[\int_{X_2} \frac{k_2^2(x;x^{(1)'},x^{(2)'})}{p_2(x;x^{(1)'},x^{(2)'})}\psi(x')dx^{(2)'}\right]dx^{(1)'}.$$

Thus, we obtain the following relations as the necessary conditions for the chain to be optimal and to have the uniform Markov property:

$$g(x) = a(x) + \int_{X_1} \frac{k_1^2(x;x^{(1)'})}{p_1(x;x^{(1)'})}\left[\int_{X_2} k_2(x;x^{(1)'},x^{(2)'})g^{1/2}(x')dx^{(2)'}\right]^2 dx^{(1)'},$$

$$p_2(x;x^{(1)'},x^{(2)'}) = \frac{|k_2(x;x^{(1)'},x^{(2)'})|g^{1/2}(x')}{\int_{X_2}|k_2(x;x^{(1)'},x^{(2)'})|g^{1/2}(x')dx^{(2)'}}.$$

In the case of constant signs, the existence and uniqueness (in L_∞) of the solution to the last nonlinear equation and the optimality of the corresponding uniform Markov chain were provided in [35] (see also [40]).

Let us also consider the estimate ξ. When constructing the latter, the weight factors $q(x,x')$ and the values of $h(x)$ can be estimated unbiasedly and independently (in the aggregate) by means of unbiased independent randomization of the corresponding integrals or even of auxiliary integral equations. The unbiasedness of the estimate thus randomized, i.e. the relation $\mathrm{E}\xi \equiv \varphi$, is obtained by direct averaging ξ [40]. From (1.12) we obtain for $\mathrm{E}\xi^2$:

$$\mathrm{E}\xi^2 = \mathrm{E}\tilde{h}^2 + 2h(\varphi - h) + \tilde{K}_p(\mathrm{E}\xi^2), \tag{1.26}$$

where $\tilde{K}_p$ is an integral operator with the kernel $\mathrm{E}\tilde{k}^2(x,x')/p(x,x')$ and $\mathrm{E}\tilde{h} \equiv h$, $\mathrm{E}\tilde{k} \equiv k$. Equation (1.26) is justified in the same manner as equation (1.18). Minimizing the value $\mathrm{E}\xi^2$ is here performed using formulae of the form of (1.22) and (1.23) with the kernel $[\mathrm{E}\tilde{k}^2(x,x')]^{1/2}$ instead of the kernel $|k(x,x')|$ and with the free term in (1.23) replaced by $\mathrm{E}\tilde{h}^2 + 2h(\varphi - h)$; the latter quantity is assumed nonnegative.

Note that for solving some applied problems, for example, problems of the radiation transfer theory, one has estimate linear functionals of the form:

$$(f, \varphi) = (\varphi^*, h) = \int_X \varphi^*(x)h(x)dx,$$

where $f \in L_1$ and φ^* is the solution to the conjugated equation $\varphi^* = K^*\varphi^* + f$; this equation is considered as the basic one in the radiation transfer theory.

Let the probability density $\pi(x)$ is such that $\pi(x) \neq 0$ for $f(x)\varphi(x) \neq 0$ and the point x distributed with the density $\pi(x)$. Then

$$(f, \varphi) = \mathrm{E}\zeta, \quad \zeta = \frac{f(x_0)}{\pi(x_0)}\xi_0.$$

One can easily see that the expression for ζ coincides with the standard collision estimate

$$\zeta = \sum_{m=0}^{N} Q_n h(x_n), \quad Q_0 = \frac{f(x_0)}{\pi(x_0)}$$

for the quantity (φ^*, h) (see [17,40]). According to the Bellman principle, minimizing the variance of this estimate is performed in the following way: one determines the optimal variant of the transition density $p(x, x')$, as was shown above, then one determines the corresponding function $\psi \equiv \mathrm{E}\xi^2$; the optimal density $\pi(x)$ is then determined by the formula:

$$\pi(x) = \frac{|f(x)|\psi^{1/2}(x)}{\int f(x)\psi^{1/2}(x)dx}.$$

1.4. BRANCHING CHAINS AND SOLUTIONS TO NONLINEAR EQUATIONS

When one constructs the basic unbiased estimate ξ, the use of branching Markov chains in solving integral equations is worthwhile in some cases. First and foremost this is true for integral equations describing the particle transfer with multiplication. In addition, trajectory 'splitting' can be used to decrease the variance and to make easier the calculation of the collision estimate. To solve such problems, an expression for the variance ξ^2 with 'splitting' (i.e. artificial branching) was constructed in [4] by averaging the corresponding recurrent representation. Considerations in the present paper show that such a way needs an additional justification, due to the fact that the condition $\|\mathrm{E}\xi^2\| < +\infty$ must be met (see Section 1.2). Here we give an expression for $\mathrm{E}\xi^2$ in a somewhat different form, which is more convenient in modelling physical processes of particle transfer with multiplication.

Let us represent the kernel of integral equation (1.8) and the transition density $p(x, x')$ as follows:

$$k(x, x') = \sum_{i=0}^{m} p_i(x)k_i(x, x')\nu_i(x'), \quad p(x, x') = \sum_{i=0}^{m} p_i(x)p_i(x, x'),$$

$$p_i(x) \geq 0, \quad \sum_{i=0}^{m} p_i(x) \equiv 1,$$

where $p_i(x, x')$ are the transition densities. Here $\nu_i(x')$ is an integer-valued function that also can have the zero value, which means that a trajectory terminates at the point x'.

Modelling the transition with branching is performed in the following way. A random number i of the transition probability $p_i(x, x')$ is chosen according to the probabilities $\{p_i(x)\}$, the transition $x \to x'$ is modelled, the weight is multiplied by the quantity $q_i(x, x') = k_i(x, x')/p_i(x, x')$ and then $\nu_i(x')$ independent trajctories with the obtained initial weight are constructed from the point x'. The corresponding recurrent representation for ξ is of the form:

$$\xi_0 = h_0 + q_i(x, x') \sum_{k=1}^{\nu_i(x')} \xi_1^{(k)}.$$

In this case the justification of unbiasedness (i.e. the relation $\mathrm{E}\xi \equiv \varphi$) presented in Section 1.2 is especially appropriate, since the full expression and direct averaging of the estimate ξ are more cumbersome than in the case without branching. By averaging the recurrent representation of ξ^2 we obtain the following expression for the function $\psi = \mathrm{E}\xi^2$, in the same manner as in Section 1.2

$$\begin{aligned}\psi(x) &= h(x)[2\varphi(x) - h(x)] + \int \sum_{i=1}^{m} p_i(x) \frac{k_i^2(x, x')\nu_i(x')[\nu_i(x') - 1]}{p_i(x, x')} \varphi^2(x')dx' \\ &+ \int \sum_{i=1}^{m} p_i(x) \frac{k_i^2(x, x')\nu_i(x')}{p_i(x, x')} \varphi(x')dx.\end{aligned}$$

As in Section 1.2. apart from the condition $\rho(K_1) < 1$ we need here the condition $\rho(K_p) < 1$, where the kernel of the operator K_p is

$$\sum_{i=1}^{m} \frac{k_i^2(x, x')\nu_i(x')}{p_i(x, x')} . \tag{1.27}$$

It can be seen from expression (1.27) that for the problem with nonnegative functional elements the operator K_p coincides with K when $p_i(x, x') \equiv k_i(x, x')$. Therefore, if $\rho(K) < 1$, then the variance of direct modelling (i.e. estimates with $Q_n \equiv 1$) is finite. Note that for a physical process of particle transfer with the branching the relation $\rho(K) < 1$ implies that the system is undercritical. In particular, for an undercritical system it is finite, if one considers $h(x)$ as the probability of particle absorption at a point x.

1.5. COST OF VARIOUS ALGORITHMS FOR SOLVING INTEGRAL EQUATIONS

This section carries out minimization and a comparative analysis of the cost of a number of Monte Carlo algorithms for solving multidimensional integral equations of the second kind. To construct a functional estimate by the Monte Carlo method, we either compute a series of integrals of the solution (a 'histogram'-type estimate) or approximate values of the solution at the nodes of the rectangular grid by using local estimates. Linear interpolation is considered as the main method of prolongation of the grid function obtained in this way. We can note that the technique developed can be extended without any difficulties to other prolongation techniques by using finite elements of higher order of smoothness.

Let us consider the integral equation of the second kind

$$f(x) = \int_X k(x', x) f(x') dx' + \psi(x) \quad \text{or} \quad f = Kf + \psi. \tag{1.28}$$

Here, X is a domain of the k-dimensional Euclidean space,

$$K \in [L_1(X) \to L_1(X)], \quad \psi \in L_1(X).$$

It is assumed that the spectral radius $\lambda(K)$ obeys the relation $\lambda(K) < 1$ and, moreover, $\lambda(K_1) < 1$, where K_1 is an operator with the kernel $|k(x', x)|$. It provides (see [39]) conditions for the standard estimates of the Monte Carlo method for linear functionals of the form $J_h = (f, h)$, where $h \in L_\infty(x)$, to be unbiased. To construct these estimates, we define in X a terminating Markov chain $x_0, ..., x_N$ with the initial density $p_0(x)$ and the transition density $p(x', x)$, and, additionally, the quantity $g(x') = 1 - \int p(x', x) dx$ is considered as a terminating probability at the point x'. We also introduce an auxiliary random weight Q_n by the formulae

$$Q_0 = \frac{\psi(x_0)}{p_0(x_0)}, \quad Q_n = Q_{n-1} \frac{k(x_{n-1}, x_n)}{p(x_{n-1}, x_n)}, \quad n = 1, 2, \ldots.$$

For standard estimate we choose the 'collision' estimate

$$\xi = \sum_{n=0}^{N} Q_n h(x_n).$$

If the sufficiently clear additional conditions for the estimate to be unbiased (see [39]) are satisfied, the following equality is valid:

$$\mathrm{E}\xi = (f, h) = \int_X f(x) h(x) dx = J_h.$$

This equality is the basis for construction of Monte Carlo algorithms which consist in the simulation of the trajectories of the Markov chain many times, with the arithmetic mean of the sampled values ξ used as the estimate of the functional J_h. The error of such estimate is probabilistic and defined by the quantity [39]

$$\mathrm{E}\xi^2 = (\chi, h[2\varphi^* - h]), \tag{1.29}$$

where φ^* is the solution to the conjugate equation $\varphi^* = K^* \varphi^* + h$, and the auxiliary function χ satisfies the following integral equation:

$$\chi(x) = \int_X \frac{k^2(x', x)}{p(x', x)} \chi(x') dx' + \frac{\psi^2(x)}{p_0(x)} . \tag{1.30}$$

Assume that $\rho(K_p) < 1$, where K_p is the integral operator in Eq. (1.30). This inequality is definitely valid for direct simulation which is possible if the kernel $k(x', x)$ is substochastic, i.e. $k(x', x) > 0$, $\int k(x', x) dx \leq 1$, and $\psi(x)$ is the probabilistic density. For direct simulation we have $p(x', x) \equiv k(x', x)$, $p_0(x) \equiv \psi(x)$, and hence, $K_p \equiv K$, $\chi(x) \equiv f(x)$.

To construct the histogram-type estimate of the solution, we use functions $\{h_i\}$ of the form

$$h_i(x) = \begin{cases} 1/\operatorname{mes}(D_i), & x \in D_i, \\ 0, & x \in D_i \end{cases}$$

for a partition $\{D_i\}_{i=1,2,...,n}$ of the domain X. Denote by ξ_i the corresponding 'collision' estimates. Introduce the notation

$$\Delta_n = \max_i \operatorname{mes} D_i \ .$$

Expression of Eq. (1.29) directly implies, for example, that the relation $E\xi^2 = O(\Delta_n^{-1})$ is valid for $\chi(x) < c < +\infty$ if $\Delta_n \to 0$, and the same time we have

$$E\xi_i \equiv f(x_i) = O(1)$$

, where x_i is a point of the subdomain D_i if we additionally assume the continuity of $f(x)$ in D_i, $i = 1, ..., n$. It is sufficient therefore to consider quantities $E\xi_i^2$ instead of the variances $D\xi = E\xi_i^2 - (E\xi_i)^2$ for $\Delta_n \to 0$. Besides, if the assumptions on the properties of the kernel are sufficiently weak, for example, if $k(x', x)$ is continuous over x' in D_i, $i = 1, 2, ..., n$, we have

$$h_i(2\varphi_i^* - h_i) \sim h_i^2, \quad \Delta_n \to 0.$$

Hence, we have

$$D\xi_i \sim (\chi, h_i^2) = \chi(x_i)[\operatorname{mes} D_i]^{-1}. \tag{1.31}$$

where x_i is a point of the subdomain D_i under the assumption that $\chi(x)$ is continuous in D_i.

Another technique of global estimation of the solution $f(x)$ consists in using so-called local estimates (see, for example, [39]) which are obtained immediately after consideration of original equation (1.28) in the form

$$f(x) = (\varphi, h_x) + \psi(x),$$

where $h_x(x') = k(x', x)$. Hence,

$$f(x) = E\left[\sum_{n=0}^{N} Q_n k(x_n, x)\right] + \psi(x) = E\xi_{0x} \ .$$

The last expression makes it possible to simultaneously estimate $f(x)$ at many points. Note that the computation of quantities $k(x_n, x)$ is usually rather costly, and when the global estimation of the solution is carried out the total cost can be considered to be proportional to the number of points at which the solution is computed. On the other hand, the variance of the local estimate is independent of the step size of the grid used unlike that of the histogram-type estimate. The comparison in detail of the costs of local and non-local estimation of the entire solution is therefore of particular interest.

The estimate of the solution at a given point can be also obtained by the 'conjugate walks' method using the expression

$$f(x) = \psi(x) + \mathrm{E}\sum_{n=1}^{N} Q_n^* \psi(x_n) = \mathrm{E}\xi_{1x}$$

assuming the original equation to be conjugate to the equation $\varphi^* = K^*\varphi^* + h$. Here, $x_0 \equiv x$, and the weights Q_n^* are computed by the formula

$$Q_n^* = Q_{n-1}^* \frac{k(x_n, x_{n-1})}{p(x_{n-1}, x_n)}, \quad Q_0 = 1, \quad n = 1, 2, \dots .$$

The disadvantage of the conjugate walks method is that the last relation makes it possible to estimate the solution only at the point x, i.e. we should simulate different trajectories for different points x. In addition, it is sometimes difficult here to choose a suitable Markov chain as the transposed kernel $k(x', x)$ is in general not substochastic even in probabilistic problems.

Let us carry out now a comparative analysis of the variances of the estimates considered above assuming for simplicity that all the functions are nonnegative and sufficiently regular at the point x. Note first of all that the local estimate can be presented as

$$h_x(x') = k(x', x) = [K^*\delta_x](x'), \quad \delta_x(x') = \delta(x' - x).$$

Therefore, we have

$$\begin{aligned} \mathrm{E}\xi_{0x}^2 &= \left(\chi, h_x[2\varphi_x^* - h_x]\right) = \left(\chi[2\varphi_x^* - h_x], h_x\right) \\ &= K\left(\chi[2\varphi_x^* - h_x]\right)(x) \geq K(\chi h_x)(x) = \int k^2(x', x)\chi(x')dx'. \end{aligned}$$

As was shown in [39], by choosing a Markov chain to be simulated we can sometimes obtain, at least, an approximate constancy of the function $\chi(x)$ which is realized, for example, for direct simulation of radiation transport in a sufficiently homogeneous medium. In this case, we have

$$\mathrm{E}\xi_{0x}^2 \geq \max_x \int k^2(x', x)dx' \tag{1.32}$$

and the quantity $\mathrm{E}\xi_{0x}^2$ can be essentially superior to the first factor in expression of Eq. (1.31) for variances of the histogram-type estimate.

For the conjugate walks method we have (see Section 1.3)

$$\mathrm{E}\xi_{1x}^2 = \psi(x)[2f(x) - \psi(x)] + \int \frac{k^2(x', x)}{p(x', x)} \mathrm{E}\xi_{1x}^2 \, . \tag{1.33}$$

It is simple to compare Eq. (1.33) with Eq. (1.31) for the case of direct simulation where we have

$$\chi(x) \equiv f(x),$$

with taking into account the inequality

$$\mathrm{E}\xi_{1x}^2 \geq f^2(x).$$

Thus, the quantity $\mathrm{E}\xi_{1x}^2$ can be essentially superior to the first factor in Eq. (1.31) in systems with a large effective number of 'collisions'.

Let us consider first the cost of histogram method. Assume that the domain X of the k-dimensional Euclidean space is partitioned into elementary subdomains $\{D_i\}$ by the rectangular grid with the step size h in each of the axes. In the histogram method the solution within the elementary subdomain is replaced with a mean integral value which is estimated as has been shown in this section. Denote by $f(x)$ an approximate solution obtained in this way. Assuming the solution to be piecewise-smooth we have

$$|f(x) - \mathrm{E}\tilde{f}(x)|^2 < ch^2,$$

where $c = \sup_x \operatorname{grad} f(x)$.

Taking into account Eq. (1.31) by analogy with Eq. (1.2) we henceforth obtain

$$\mathrm{E}\|f(x) - \tilde{f}(x)\|^2_{L_2} \lesssim ch^2 + \frac{d}{nh^k}, \tag{1.34}$$

where n is the size of the sample of trajectories of the Markov chain.

To minimize the cost of the histogram method, it is thus necessary to solve the following problem:

$$S = nt \to \min_{N,h}, \quad ch^2 + \frac{d}{nh^k} = \delta^2$$

or (substituting n from the last equality)

$$\frac{td}{(\delta^2 - ch^2)h^k} \to \min_h, \quad \delta^2 - ch^2 > 0. \tag{1.35}$$

Here, δ is the required error, and t is an average cost of one sample which can be obviously considered here to be independent of h. By direct differentiation from Eq. (1.35) we obtain

$$h^* = \left[\frac{k}{(2+k)c}\right]^{1/2} \cdot \delta\,.$$

Henceforth,

$$n^* = \frac{(2+k)d}{2\delta^2 h^{*k}} = \frac{d(2+k)\left[(2+k)c/k\right]^{k/2}}{2\delta^{2+k}}\,.$$

Hence, we have here

$$S^* = \frac{d(2+k)\left[(2+k)c/k\right]^{k/2}}{2\delta^{2+k}}\,.$$

Let us consider now a simple but practically efficient modification of the histogram method, which yields an estimate called in statistics as 'frequency polygon'. The frequency polygon of $\tilde{f}^{(1)}(x)$ at midpoints D_i coincides with the histogram $\tilde{f}(x)$ and at the points of the domain X it is obtained by linear interpolation in such a way that

$$\left[f(x) - \mathrm{E}\tilde{f}^{(1)}(x)\right]^2 \le c^{(1)}h^4, \quad \mathrm{E}\|f(x) - \tilde{f}^{(1)}(x)\|^2_{L_2} \le c^{(1)}h^4 + \frac{d}{nh^k}\,.$$

The coefficient $c^{(1)}$ can be estimated via the upper bound of the second-order derivatives depending on the linear prolongation technique.

To minimize the cost, in this case we have to solve the following problem:

$$S_1 = nt \to \min_{n,h}, \quad c^{(1)}h^4 + \frac{d}{nh^k} = \delta^2.$$

In the same way as for the histogram method we obtain here

$$h_1^* = \left[\frac{k}{c^{(1)}(4+k)}\right]^{1/4} \delta^{1/2},$$

$$S_1^* = tn_1^* = t\frac{d(4+k)\left[(4+k)c^{(1)}/k\right]^{k/4}}{4\delta^{2+k/2}}.$$

Thus, for multidimensional problems the frequency polygon is essentially more efficient than the histogram if we have an optimal relation between n and h.

Let us assume now that the values of $f(x)$ at the grid nodes are computed by using the local estimate ξ_{0x} with the further piecewise-linear prolongation as in the case of the frequency polygon. Assume that $D\xi_{0x} < +\infty$. Then, with the previous remark on the asymptotics of the computational cost of local estimates taken into account, the cost minimization problem is of the form

$$S_l = nt_l h^{-k} \to \min_{n,h}, \quad c^{(1)}h^4 + \frac{d_l}{n} = \delta^2. \tag{1.36}$$

Here, $t_l = \text{mes}(X) \cdot t_{0l}$, where t_{0l} is the cost of computation of the function $h_x(x') = k(x', x)$ for one value of x, and

$$d_l = \max_x D\xi_{0x}.$$

The minimization problem considered can be reduced to the previous one by making the replacement $m = nh^{-k}$. As a result, we have

$$h_l^* = \left[\frac{k}{(4+k)c^{(1)}}\right]^{1/4} \delta^{1/2}, \quad n_l^* = \frac{d_l(4+k)}{4\delta^2},$$

$$S_l^* = t_l\frac{d_l(4+k)[(4+k)c^{(1)}/k]^{k/4}}{4\delta^{2+k/2}}$$

that is the expression S_l^* can be obtained from the expression S_1^* by making the following replacement: $t \to t_l$, $\quad d \to d_l$. This enables us to effectively compare in practice the costs of the frequency polygon method and the local estimates method with an optimal choice of the parameters n and h. First, it is obvious that, as a rule, $t \approx t_l$. For example, in solving the transport theory problems by the direct simulation t is an average time of simulation of the particle trajectory, i.e. $t = \text{E}(N)t_0$, t_0 is the time of simulation of an elementary transition corresponding to the substochastic kernel. Sometimes, we have $t_0 \approx t_{0l}$ and

$$\frac{t}{t_l} \approx 1/\text{mes}(X).$$

On the other hand, relations of Eq. (1.31) and Eq. (1.32) show that if $\chi(x) \approx \text{const}$, then we have

$$\frac{d_l}{d} > \max_x \int_X k^2(x', x)dx'.$$

Note that the relation

$$\frac{n_1^*}{n_l^*} = O(\delta^{-k/2})$$

can be considered to be in favour of the histogram method as the greater number of trajectories simulated makes it possible to estimate with a greater precision certain special functionals, root-mean-square errors and also, for example, probabilistic characteristics of solutions to problems with random parameters when double randomization is used. Also note that the optimal relation between the parameters n and h in the histogram method can be estimated in a rather simple way. On the other hand, with the local estimate used, the values of $f(x)$ are estimated in a more dependent way, this making it possible to estimate with a greater precision, for example, the quantity $\arg\sup f(x)$. In addition, when using local estimates, we can utilize in principle a smoother prolongation of order $m \ll h^{-k}\,\text{mes}(X)$. The corresponding cost minimization problem is of the form

$$S_l^{(m)} = nt_l h^{-k} \to \min_{n,h}, \quad c^{(m)}h^{2m} + \frac{d_l}{n} = \delta^2.$$

This problem can be reduced to the previous one, i.e. to the case of $m = 2$ by making the following replacement: $h \to h_0^{2/m}, \quad k \to 2k/m$. As a result, we obtain

$$h_l^{(m)} = \left[\frac{k}{(2m+k)c^{(m)}}\right]^{1/2m} \delta^{1/m}, \quad n_l^{(m)} = \frac{d_l(2m+k)}{2m\delta^2},$$

$$S_l^{(m)} = t_l \frac{d_l(2m+k)\left[(2m+k)c^{(m)}/k\right]^{k/(2m)}}{2m\delta^{2+k/m}}.$$

The minimization of the estimation cost for the solution $f(x)$ by the 'conjugate walks' method is obviously carried out in the same way as for the local estimates method and leads to the same expressions with the following replacement: $t_l \to t', \quad d_l \to d'$. The quantity t' means the average cost of simulation of the conjugate trajectory and can be close to the average cost of simulation of a trajectory in the histogram method, i.e. to t. But the quantity d' can be essentially superior to the corresponding coefficient d in the histogram method. This is implied, for example, by estimates of Eq. (1.31) and Eq. (1.33).

The local estimates used for the solution of the transport theory problems, as a rule, have infinite variances (see Section 5.1) because of the fact that

$$\int_X k^2(x', x)dx' = +\infty.$$

We can overcome this difficulty by carrying out an 'ε-cut' of the kernel when computing the local estimate, i.e. we can assume that

$$h_x(x') = \begin{cases} k(x', x), & |x' - x| > \varepsilon, \\ 0, & |x' - x| \le \varepsilon. \end{cases} \tag{1.37}$$

In this case, we have $D\xi_{0x} \lesssim d\varphi(\varepsilon)$. Assume that the corresponding bias in the estimate of $f(x)$ is of order ε (otherwise we can find a suitable change for this parameter). For simplicity, we consider the linear prolongation of local estimates taking into account the fact that the transition to a more accurate prolongation is carried out automatically.

We thus arrive at the following cost minimization problem for the biased local estimate of the solution:

$$S_\varepsilon = t_l n h^{-k} \to \min_{n,h,\varepsilon}, \quad c^{(1)}h^4 + b\varepsilon^2 + \frac{d_0\varphi(\varepsilon)}{n} = \delta^2.$$

This problem can be first reduced to Eq. (1.36) by making the following replacements:

$$d_l \to d_0\varphi(\varepsilon), \quad \delta^2 \to \delta^2 - b\varepsilon^2.$$

As a result, we obtain

$$h_\varepsilon^* = \left[\frac{k}{(4+k)c^{(1)}}\right]^{1/4}(\delta^2 - b\varepsilon^2)^{1/4}, \quad n_\varepsilon^* = \frac{d_0\varphi(\varepsilon)(4+k)}{4(\delta^2 - b\varepsilon^2)},$$
$$S_\varepsilon^* = t_l \frac{d_0\varphi(\varepsilon)(4+k)[(4+k)c^{(1)}/k]^{k/4}}{4(\delta^2 - b\varepsilon^2)^{1+k/4}}. \tag{1.38}$$

Then we have to carry out minimization of S_ε^* over ε. The equation $\partial S_\varepsilon^*/\partial\varepsilon = 0$ can be reduced to the form

$$\delta^2 - b\varepsilon^2 = -\frac{4+k}{2} b \frac{\varepsilon\varphi(\varepsilon)}{\varphi'(\varepsilon)}. \tag{1.39}$$

Substituting the solution to this equation into h_ε^*, n_ε^*, and S_ε^*, we obtain optimal values of all the quantities.

Let us briefly discuss the transition to a higher order prolongation in this case. To this end, we should replace k with $2k/m$ in the expression S^*. It is obvious that such replacement does not affect the order of solution of equation (1.39) (it is valid for particular cases $\varphi(\varepsilon)$ important in practice which are to be considered henceforth). The main effect here thus consists in the replacement of the exponent in the denominator of the expression for the optimal cost: $1 + k/4 \to 1 + k/(2m)$. Note that as compared with the case of finite variance this effect is less important due to the presence of the factor $\varphi(\varepsilon)$. Now let us separately consider two practically important cases of $\varphi(\varepsilon)$.

Case study 1. In this case, we have $\varphi(\varepsilon) = 1/\varepsilon$. Here equation (1.39) takes the form

$$\delta^2 - b\varepsilon^2 = \frac{4+k}{2} b\varepsilon^2.$$

Hence, we obtain

$$\varepsilon^* = \delta\left[\frac{2}{b(6+k)}\right]^{1/2}, \quad \delta^2 - b\varepsilon^{*2} = \frac{4+k}{6+k}\delta^2,$$
$$n_{\varepsilon*}^* = \frac{d_0\sqrt{b}(6+k)^{3/2}}{4\sqrt{2}}\delta^{-3}, \quad h_{\varepsilon*}^* = \left[\frac{k(6+k)}{(4+k)^2c^{(1)}}\right]^{1/4}\delta^{-1/2},$$
$$S_{\varepsilon*}^* = t_l \frac{d_0\sqrt{b}(c^{(1)}/k)^{k/4}(6+k)^{3/2+k/4}}{4\sqrt{2}}\delta^{-3-k/2}.$$

The comparison of the quantities $S_{\varepsilon*}^*$ and S_1^* shows that in this case the local estimation is essentially inferior to the histogram method as far as the cost is concerned. It is also interesting to find out the role of the simple equalization of separate terms in the error estimate in this case. Denote by $S_\varepsilon^{(0)}$ the corresponding cost. The arguments show that

$$\frac{S_\varepsilon^{(0)}}{S_{\varepsilon^*}^*} = \frac{k^{k/4}}{2^{k/4-1}} \,.$$

This relation is, for example, equal to 2.71 for the case of $k = 3$ important in practice.

Case study 2. In this case, we have $\varphi(\varepsilon) = |\ln \varepsilon| = -\ln \varepsilon$. Here, equation (1.39) is of the form

$$\delta^2 - b\varepsilon^2 = -\frac{4+k}{2} b\varepsilon^2 \ln \varepsilon. \tag{1.40}$$

Hence, we obtain the relation

$$\varepsilon^2 \left[2 + (4+k)|\ln \varepsilon|\right] = \frac{2\delta^2}{b} \,. \tag{1.41}$$

Equation (1.40) as well directly implies the following expression for the optimal cost via the optimal value $\varepsilon = \varepsilon^*$:

$$S_{\varepsilon*}^* = t_l \frac{d_0(4+k)[(4+k)c^{(1)}k]^{k/4}}{4\left[(4+k)b\varepsilon^{*2}/2\right]^{1+k/4} |\ln \varepsilon^*|^{k/4}} \,.$$

Consider now the problem of practical computation of the solution to equation (1.41). First, note that Eq. (1.41) reduces asymptotically for $\delta \to 0$ to the equation

$$\varepsilon^2 |\ln \varepsilon| = \frac{2\delta^2}{b(4+k)} \,. \tag{1.42}$$

Equation (1.42) implies that asymptotically $\varepsilon^* < \delta\sqrt{2}/\sqrt{b(4+k)}$. It is now obvious that the value of the solution also increases in replacing $|\ln \varepsilon|$ with $|\ln \delta|$ in Eq. (1.42), i.e. asymptotically

$$\varepsilon^* < \frac{\delta\sqrt{2}}{\left[|\ln \delta| b(4+k)\right]^{1/2}} = \varepsilon_1 \,.$$

Since the denominator of the expression $S_{\varepsilon^*}^*$ tends to zero for $\varepsilon^* \to 0$, then if ε^* is replaced with a greater quantity ε_1, in the asymptotics there should be a decrease in the value of $S_{\varepsilon^*}^*$. Therefore, asymptotically we have

$$S_{\varepsilon^*}^* > t_l \frac{d_0(4+k)\left[(4+k)c^{(1)}k\right]^{1/4}}{4} \frac{|\ln \delta|}{\delta^{2+k/2}} = S_0^* \,.$$

On the other hand, if we substitute the non-optimal value $\varepsilon = \varepsilon_1$ into Eq. (1.39), we obviously obtain the upper bound for $S_{\varepsilon^*}^*$, i.e. we have $S_0^* < S_{\varepsilon^*}^* < S_{\varepsilon_1}^*$. It is not difficult to verify that $S_{\varepsilon_1}^*$ asymptotically coincides with S_0^*.

We have thus proved the following

Theorem 1.3. *For $\varphi(\varepsilon) = |\ln \varepsilon|$ the value*

$$\varepsilon_1 = \frac{\delta\sqrt{2}}{\left[|\ln\delta| b(4+k)\right]^{1/2}}$$

is asymptotically optimal for $\delta \to 0$. The corresponding value of the cost asymptotically coincides with S_0^*. □

Taking into account the relation $\varepsilon_1^2/\delta^2 \to 0$ we also have the following asymptotic expressions:

$$h_{\varepsilon_1}^* = \left[\frac{k}{(4+k)c^{(1)}}\right]^{1/4}\delta^{1/2}, \quad n_{\varepsilon_1}^* = \frac{d_0(4+k)}{4\delta^2}|\ln\delta|.$$

Compare now the optimal cost of the 'frequency polygon' method with that of the local estimation logarithmically biased. We have

$$\frac{S_{\varepsilon^*}^*}{S_1^*} = \frac{t_l d_0}{td}|\ln\delta|.$$

If $\mathrm{mes}(X) \ll 1$ then the last quotient can be essentially less than unity and the logarithmically biased estimation for not too small δ can be preferable that the frequency polygon method especially in those cases where the smoothness of the estimate of $f(x)$ is useful.

Let us make a remark on the estimate of the square biasedness $c^{(1)}h^4 + b\varepsilon^2$ used in this section, due to the form of which we have managed to obtain the results in the complete form. Another estimate

$$(ch^2 + b_1\varepsilon)^2 \le 2(c^2h^4 + b_1^2\varepsilon^2)$$

is standard. The estimate however is obviously less accurate if the considered forms of biasedness have opposite signs, which is the frequent case in transport theory problems due to an exponential (over the space) decrease in radiation intensity. Otherwise we can introduce a factor q in the estimate used, and also $1 \le q \le 2$. This, obviously, leads to the following replacement: $\delta^2 \to \delta^2/q$, $d_0 \to d_0/q$ which practically does not effect the optimal values of h and n.

In this Section there were considered the global estimates with respect to the metric $L_2(X)$. It is possible to obtain the similar results for the metric $C(X)$ by using the theorems of imbedding of the space $W_2^l(X)$ into the space $C(D)$, provided $2l > k$. When using local estimates or the 'conjugate walks' method these results are the same as in Section 1.1 with specific constants. If the proper derivatives of the kernel exist, then the cost of local estimates method is similar to the cost of dependent simulating the 'conjugate walks'. The similar problem for the 'frequency polygon' will be considered in Section 5.7 in comparison with the 'free-path estimate' method.

1.6. MONTE CARLO METHODS FOR SOLVING PROBLEMS WITH STOCHASTIC PARAMETERS

Various examples introducing additional randomness for constructing effective simulation algorithms can be found in the literature devoted to the Monte Carlo methods (see, for example, [17]). This section is concerned with the optimization of randomized algorithms for estimating probabilistic characteristics of equations with random parameters. In this connection, randomized models for random fields are suggested in the Appendix.

Randomized estimation for the statistical moments of the solution is presented be low. Assume a linear functional equation $L\phi = f$ to be solved by the Monte Carlo method on the basis of simulation of a stochastic process. (Denote the trajectories of this process by ω). This means that random variables $\xi_k(\omega)$ are constructed so that

$$\mathrm{M}\xi_k(\omega) = J_k, \quad k = 1, 2, ..., m,$$

where J_k are the functionals of ϕ to be evaluated (M denotes the mathematical expectation).

Let the operator L and the function f depend on a random field σ (for example, a random medium in transfer theory, random force in elasticity theory, etc.). Also,

$$\xi_k = \xi_k(\omega, \sigma), \quad J_k = J_k(\sigma)$$

and

$$\mathrm{M}[\xi_k(\omega, \sigma)|\sigma] = J_k(\sigma)$$

, where the variables ω and σ are generally not independent.

Consider the problem of evaluating the quantities

$$J_k = \mathrm{E}J_k(\sigma), \quad R_{kj} = \mathrm{E}[J_k(\sigma)J_j(\sigma)], \quad k, j = 1, ..., m,$$

where E denotes the mathematical expectation with respect to the distribution of σ.

The following obvious method is known for evaluating these mathematical expectations. First, realizations of σ are constructed: then the equation $L_\sigma \phi_\sigma = f_\sigma$ is solved precisely enough for each realization by a numerical or an analytical technique. Finally, statistical estimates of the desired quantities are calculated. However, this approach fails for complicated multi-dimensional problems because the computational cost of an explicit solution of the equation considered is too high. Therefore, it is useful to apply sometimes a method of 'double randomization'. In our case, this technique follows from the following relations:

$$\begin{aligned} \mathrm{E}J_k(\sigma) &= \mathrm{EM}\xi_k(\omega, \sigma) = \mathrm{M}_{(\omega,\sigma)}\xi_k(\omega, \sigma), \\ \mathrm{E}[J_k(\sigma)J_j(\sigma)] &= \mathrm{M}_{(\omega_1,\omega_2,\sigma)}[\xi_k(\omega_1, \sigma)\xi_j(\omega_2, \sigma)], \end{aligned} \tag{1.43}$$

where ω_1 and ω_2 are conditionally independent trajectories, constructed for one fixed realization of σ and the subscript of the expectation symbol indicates the distribution to which it corresponds. Relations of Eq. (1.43) show that to estimate the quantities J_k it is sufficient to construct only one trajectory for a fixed σ, while the estimation of the quantities R_{kj} requires two conditionally independent trajectories. To optimize the randomization technique, it is natural to use the 'splitting method' (see, for example, [40]). In this method, the quantities J_k are estimated as follows. First, one constructs n conditionally independent trajectories (i.e. a vector $\omega = (\omega_1, ..., \omega_n)$, with σ fixed), and then a random variable

$$\zeta_k^{(n)}(\omega, \sigma) = \frac{1}{n}\sum_{i=1}^{n} \xi_k(\omega_i, \sigma)$$

is used instead of $\xi_k(\omega, \sigma)$. The optimal value of n is calculated by the formula (see, for example, [40])

$$n = \sqrt{\frac{a_2}{a_1} \cdot \frac{t_1}{t_2}}\,, \tag{1.44}$$

where

$$a_1 = \mathrm{E}(\mathrm{M}\xi_k)^2 - J_k^2, \quad a_2 = \mathrm{E}\mathrm{D}\xi_k\,,$$

t_1 is the average computing time for a fixed realization of σ, and t_2 is the average computing time for a fixed realization of ω.

It is difficult to evaluate the quantities a_1, t_1, t_2 directly. However, one can obtain statistical estimates of the variances and the computing times for two values of the splitting parameter, n_1, n_2, and then solve the corresponding system of linear equations, i.e. make use of the following equalities:

$$\begin{aligned} a_1 &= \frac{1}{n_2 - n_1}(n_2 \mathrm{D}\zeta_k^{(n_2)} - n_1 \mathrm{D}\zeta_k^{(n_1)}), \qquad a_2 = \left[\frac{n_1 n_2}{n_2 - n_1}\right](\mathrm{D}\zeta_k^{(n_1)} - D\zeta_k^{(n_2)})\,, \\ t_1 &= \frac{n_2 t^{(n_1)} - n_1 t^{(n_2)}}{n_2 - n_1}, \qquad t_2 = \frac{t^{(n_1)} - t^{(n_2)}}{n_1 - n_2}\,. \end{aligned} \tag{1.45}$$

Here, it is useful to correlate the samples of $\zeta_k^{(n_1)}$ and $\zeta_k^{(n_2)}$.

Let us consider now the optimization of the randomized estimate of the quantity R_{kj} using the splitting technique. In this case, it is natural to use the random variable

$$\rho_{kj}^{(n)} = \frac{\sum_{i=1}^{n} \sum_{t=i+1}^{n} [\xi_k(\omega_i, \sigma)\xi_j(\omega_t, \sigma)]}{n(n-1)/2}\,, \qquad \mathrm{M}\rho_{kj}^{(n)} = R_{kj}\,.$$

Note that instead of $\rho_{kj}^{(n)}$ it is possible to use another quantity which is numerically equivalent to $\rho_{kj}^{(n)}$, but more convenient for calculations. Indeed, the following easily verified relation for the covariances holds:

$$K[\zeta_k^{(n)}(\omega, \sigma), \zeta_j^{(n)}(\omega, \sigma)] = K[J_k(\sigma), J_j(\sigma)] + \frac{1}{n}\mathrm{E}K[\xi_k, \xi_j].$$

Taking into account the analogous relation for $n = 1$, we obtain from this system of equations that

$$K_{jk} = K[J_k(\sigma), J_j(\sigma)] = \frac{nK[\zeta_k^{(n)}, \zeta_j^{(n)}] - K[\zeta_k^{(1)}, \zeta_j^{(1)}]}{n-1}\,. \tag{1.46}$$

It is not difficult to verify that substituting the statistical estimates for the appropriate covariances in the right-hand side of Eq. (1.46) yields an estimate, $\tilde{K}_{jk}^{(n)}$, of the quantity K_{jk} which coincides numerically with the estimate obtained on the basis of $\rho_{kj}^{(n)}$.

When Eq. (1.46) is used, the average computing time is given, as in the standard splitting technique, by the formula

$$T^{(n)} = T_1 + nT_2\,.$$

The explicit formula for the variance $\mathrm{D}\rho_{kj}^{(n)}$ is very cumbersome. However, it is clear that for sufficiently large n the following approximate relation holds:

$$\mathrm{D}K_{kj} \approx A_1 + \frac{A_2}{n} .$$

Consequently, the splitting technique, when used for estimating K_{jk}, also can be optimized by a formula of type of Eq. (1.44) or of Eq. (1.45). It should be noted that we do not consider here the optimization of the choice (in some sense) of n for estimating all the quantities K_{jk}, $k = 1, ..., m$, simultaneously.

Assume now the random field σ to be entirely defined by the finite number of the random parameters $\sigma_1, ..., \sigma_r$. In addition, assume the dependence of the functionals $J_k(\sigma)$ on σ_i, $i = 1, ..., r$ to be sufficiently linear (this assumption can be made in the neighbourhood of the point $\sigma^{(0)} = \mathrm{M}\sigma$; the probability to hit this neighbourhood is close to unity). Then the correlations of the vector $J = J_k(\sigma)$ and the autocorrelations of the vectors σ and J can be evaluated by linearizing $J(\sigma)$, using Monte Carlo calculations of the estimates of the derivatives $\partial J_k/\partial\sigma_i$. The corresponding numerical technique is given in the transfer theory [28] for piecewise-constant random cross-section σ.

Consider optimization of the double randomization method for calculating functionals of the form

$$J = \mathrm{E}J(\sigma) = \mathrm{E}\mathrm{M}\xi(\omega, \sigma)$$

taking into account the cost of the approximate simulation of the field σ. Let σ_m be an approximate model for the field σ (see Appendix), and assume that

$$(J_m - J)^2 \lesssim cm^{-\alpha}, \quad J_m = \mathrm{E}J(\sigma_m), \quad \alpha > 0.$$

Here, the symbol $\lesssim$ means that the estimate is sufficiently exact. Inequalities of this type can be obtained using *a priori* information of the solution on the basis of estimates of the corresponding norms of the random function $\sigma - \sigma_m$. Assume also the variance

$$d = \mathrm{D}\xi(\omega, \sigma_m) = \mathrm{E}\mathrm{M}[\xi(\omega, \sigma_m) - J_m]^2$$

to be independent of m. The probabilistic error of the estimate for $J(\omega, \sigma_m)$ is defined by the quantity

$$\mathrm{E}\mathrm{M}[\xi(\omega, \sigma_m) - J]^2 = d + (J_m - J)^2.$$

For $\xi_N(\omega, \sigma_m)$, i.e. for the average value of N independent realization of $\xi(\omega, \sigma_m)$, we have

$$\mathrm{E}\mathrm{M}[\xi_N(\omega, \sigma_m) - J]^2 \lesssim \frac{d}{N} + cm^{-\alpha}.$$

Let us measure the cost of an algorithm by the average number of operations required to obtain the result to within probabilistic error ε. Assume the average cost of simulation of $\xi(\omega, \sigma)$ for $\sigma = \sigma_m$ to be given by $c_0 m^\beta$, $\beta > 0$. Note that in the algorithms of Section 1.7 β is apparently equal to unity. To optimize the randomized algorithm of estimation of the quantity J, it is thus necessary to solve the following conditional minimization problem:

$$\min_{N,m} \left\{ Nm^\beta, \frac{d}{N} + cm^{-\alpha} = \varepsilon^2 \right\}. \tag{1.47}$$

From Eq. (1.47) we have

$$m = \left[\frac{1}{c}\left(\varepsilon^2 - \frac{d}{N}\right)\right]^{-1/\alpha},$$

i.e. it is necessary to minimize the quantity

$$f(N) \equiv N\left(\varepsilon^2 - \frac{d}{N}\right)^{-\beta/\alpha}.$$

The solution of the equation $f'(N) = 0$ has the form:

$$N = (d + \beta/\alpha)\varepsilon^{-2} \quad \text{or} \quad \varepsilon^2 = \frac{d + \beta/\alpha}{N}.$$

From Eq. (1.47), we now have

$$\frac{d}{N} + cm^{-\alpha} = \frac{d + \beta/\alpha}{N},$$

i.e.

$$N = \frac{\beta}{\alpha c} m^{\alpha}.$$

The main difficulty here when solving a concrete problem is to obtain sufficiently exact values of c and α. It is interesting to note that the last expression holds also if the splitting technique is used, i.e. when $\xi(\omega, \sigma)$ is replaced by $\zeta^{(n)}(\omega, \sigma)$, provided that the average cost of one sampling of $\zeta^{(n)}(\omega, \sigma)$ for $\sigma = \sigma_m$ is also proportional to m^{β}.

Chapter 2

Estimation of derivatives

2.1. VECTOR MONTE CARLO ALGORITHMS

Vector Monte Carlo algorithms are used to solve systems of integral equations. The distinguishing feature of the vector algorithm is that its 'weight' appears to be a matrix weight. This matrix weight is multiplied by the kernel matrix of the system of integral equations divided by a transition density function after each transition in the Markov chain simulation. The following vector Monte Carlo algorithms are well known: an algorithm for solving the system of transfer equations with polarization taken into account [28], a vector algorithm for solving multigroup transfer equations [52], a Monte Carlo technique combined with the finite sum method [40] and a vector Monte Carlo method for solving metaharmonic equation [47]. We also mention the weight algorithms, so-called depending sampling methods (see, for example, [40]), which are used to solve equations whose kernel contains a parameter. So a number of computational problems can be reduced to the solution of a system of integral equations of second kind. By introducing a suitable discrete-continuous measure of integration, we can write such a system in the form of a single integral equation, and this allows us to use standard algorithms of the Monte Carlo method. Sometimes, however, it is more expedient to make use of the matrix structure of the system and to solve the problem by the Monte Carlo method with vector weights. A general formulation of vector algorithms of this type is given below. Consider the system of integral equations

$$\varphi_i(x) = \sum_{j=1}^{m} \int_X k_{ij}(x,y)\varphi_j(y)dy + h_i(x), \quad i = 1, 2, ..., m, \tag{2.1}$$

where $x, y \in X$, and the integral is taken with respect to Lebesgue measure in a finite-dimensional euclidean space X. Let us introduce the notation

$$K(x,y) = \{k_{ij}(x,y)\}_{i,j=1,2,...,m}, \quad \Phi(x) = \big(\varphi_1(x), \varphi_2(x), ..., \varphi_m(x)\big)',$$

the prime denotes the transpose, i.e. Φ is a column vector. If we regard the integral of the matrices as a matrix of integrals, we can rewrite system Eq. (2.1) in the form

$$\Phi(X) = \int_x K(x,y)\Phi(y)dy + H(x) \tag{2.2}$$

or $\Phi = \tilde{K}\Phi + H$. We will consider Eq. (2.2) in L_∞, letting

$$\|\Phi\| = \text{vrai} \sup_{i,x} |\varphi_i(x)|;$$

at the same time

$$\|\tilde{K}\| \leq \text{vrai} \sup_{i,x} \sum_j \int |k_{ij}(x,y)|dy = q.$$

Let us specify a homogeneous Markov chain $x_0 = x, x_1, \ldots, x_N$ in X by the substochastic transition density $p(x \to y) = p(x,y)$. Here the random length of the chain, which terminates upon transition from the point x with probability

$$P(x) = 1 - \int p(x,y)dy\,.$$

We will suppose that

$$p(x,y) \neq 0 \quad \text{if} \quad K(x,y)\Phi(y) \neq 0.$$

We will assume also that the spectral radius of the operator $\tilde{K}_1$ obtained from $\tilde{K}$ by the substitution $k_{ij} \to |k_{ij}|$, $i,j = 1,\ldots,m$ is less than one. Then, by using standard devices of Monte Carlo theory (e.g. [40]) we can show that

$$\Phi(x) = \mathrm{E}\xi_x, \quad \xi_x = \sum_{n=0}^{N} Q_n H(x_n), \tag{2.3}$$

$$Q_0 = \{\delta_{ij}\}_{i,j=1,2,\ldots,m}\,, \quad Q_n = Q_{n-1}\frac{K(x_{n-1},x_n)}{p(x_{n-1},x_n)}\,,$$

here

$$K(x_{n-1},x_n)/p(x_{n-1},x_n) = \{k_{ij}(x_{n-1},x_n)/p(x_{n-1},x_n)\}_{i,j=1,2,\ldots,m}\,,$$

and the Q_n can be considered as matrix weights. The Monte Carlo method is ordinarily used to estimate linear functionals of the form

$$I = (F,\Phi) = \int F'(x)\Phi(x)dx,$$

where $F'(x) = (f_1(x), ..., f_m(x))$, with

$$\|F\|_{L_1} = \sum_j \int |f_j(x)|dx < +\infty\,.$$

Let the point x_0 be distributed with probability density $\pi(x)$ such that $\pi(x) \neq 0$ if $F'(x)\Phi(x) \neq 0$. Then, obviously,

$$I = \mathrm{E}\left[\frac{F'(x_0)}{\pi(x_0)}\xi_{x_0}\right] = \mathrm{E}\sum_{n=0}^{N}\frac{F'(x_0)}{\pi(x_0)}Q_n H(x_n) = \mathrm{E}\sum_{n=0}^{N} H'(x_n)Q_n'\frac{F(x_0)}{\pi(x_0)}.$$

The random vector with weight

$$Q_n^{(')} = Q_n' F(x_0)/\pi(x_0)$$

is computed by the formula

$$Q_n^{(')} = [K'(x_{n-1}, x_n)/p(x_{n-1}, x_n)]Q_{n-1}^{(')}.$$

Precisely such a vector algorithm, corresponding to the representation $I = (\Phi^*, H)$, has been formulated in [28]. We note that in the equation $\Phi^* = \tilde{K}^*\Phi^* + F$ is taken as basic, and the weights $Q_n^{(')}$ are transformed by the untransposed matrix kernel. Setting $\zeta = F'(x_0)\xi_{x_0}/\pi(x_0)$, we have

$$\mathrm{E}\zeta^2 = \mathrm{E}[F'(x_0)\xi_{x_0}\xi'_{x_0}(x_0)/\pi^2(x_0)]\,.$$

Thus the variance of ζ is determined by the matrix-valued function $\Psi(x) = \mathrm{E}[\xi_x\xi'_x]$, which we study below. In what follows we will assume finiteness of the matrix-valued function $\Psi_1(x) = \mathrm{E}[\xi_x^{(1)}\xi_x^{(1)'}]$, where $\xi_x^{(1)}$ is obtained as a result of the substitutions $k_{ij} \to |k_{ij}|$ and $h_i \to |h_i|$. This assumption justifies the subsequent interchange of summation and integration, and also assures the finiteness of $\Psi(x)$ as in the scalar case (see Sect. 1.3). Corresponding to Eq. (2.3), we have

$$\begin{aligned}
\Psi(x) &= \mathrm{E}\sum_{n=0}^{N} Q_n H(x_n)\cdot H'(x_n)Q'_n + \mathrm{E}\sum_{n=0}^{N} Q_n H(x_n)\left[\sum_{m=n+1}^{N} Q_m H(x_m)\right]' \\
&\quad + \mathrm{E}\sum_{n=0}^{N}\left[\sum_{m=n+1}^{N} Q_m H(x_m)\right] H'(x_n)Q'_n \\
&= E_1 + \mathrm{E}\sum_{n=0}^{N} Q_n H(x_n)[Q_n(\xi_{x_n} - H(x_n))]' \\
&\quad + \mathrm{E}\sum_{n=0}^{N} [Q_n(\xi_{x_n} - H(x_n))]H'(x_n)Q'_n \\
&= E_1 + \mathrm{E}\sum_{n=0}^{N} Q_n H(x_n)[\Phi'(x_n) - H'(x_n)]Q'_n \\
&\quad + \mathrm{E}\sum_{n=0}^{N} Q_n[\Phi(x_n) - H(x_n)]H'(x_n)Q_n \\
&= \mathrm{E}\sum_{n=0}^{N} Q_n[H(x_n)\Phi'(x_n) + \Phi(x_n)H'(x_n) - H(x_n)H'(x_n)]Q'_n. \qquad (2.4)
\end{aligned}$$

Here we have used the standard conditional averaging for a fixed value of the sequence $x_0, x_1, ..., x_n$.

The expectation of the n-th term of the sum in Eq. (2.4) is expressed as follows:

$$\begin{aligned}
\int \dots \int \frac{K(x_0, x_1)}{p(x_0, x_1)} \cdots \frac{K(x_{n-1}, x_n)}{p(x_{n-1}, x_n)} G(x_n)K'(x_{n-1}, x_n)\dots \\
\dots K'(x_0, x_1)dx_0 dx_1 \dots dx_n\,,
\end{aligned} \qquad (2.5)$$

where $G = H\Phi' + \Phi H' - HH'$. It is not hard to see that Eq. (2.5) is the n-th term of the Neumann series for the equation

$$\Psi(x) = H\Phi' + \Phi H' - HH' + \int \frac{K(x,y)}{p(x,y)} \Psi(y) K'(x,y) dy. \tag{2.6}$$

We have thus proved the following

Proposition. *If $\Psi_1(x) < +\infty$, then the Neumann series for Eq. (2.6) converges and has the representation $\Psi(x) = \mathrm{E}[\xi_x \cdot \xi_x']$.* □

Now consider the question of a criterion for finiteness of the matrix-valued function $\Psi_1(x)$, which is expressed by the Neumann series for Eq. (2.6) after the substitutions $k_{ij} \to |k_{ij}|$ and $h_i \to |h_i|$. We denote the corresponding integral operator by the symbol $\tilde{K}_p^{(1)}$. With the help of a direct estimate we obtain

$$\begin{aligned} \|\tilde{K}_p^{(1)}\| &\le \operatorname*{vrai\,sup}_{i,x} \int_X \frac{\left(\sum_j |k_{ij}(x,y)|\right)^2}{p(x,y)} dy \\ &\le \left(\operatorname*{vrai\,sup}_{i,x,y} \sum_j \frac{|k_{ij}(x,y)|}{p(x,y)} \right) \sup_{i,x} \sum_j \int |k_{ij}(x,y)| dy = q_p . \end{aligned}$$

Consequently, if $q_p < 1$, then $[\xi_x \xi_x'] < +\infty$.

By looking at the n-th term of the Neumann series for Ψ_1, we can easily widen this condition in the following way: the function $\Psi_1(x)$ is finite if

$$\left(\operatorname*{vrai\,sup}_{i,x,y} \sum_j \frac{|k_{ij}(x,y)|}{p(x,y)} \right) \rho(\tilde{K}_1) < 1,$$

where $\rho(\tilde{K}_1)$ is the spectral radius of the operator $\tilde{K}_1$ obtained from $\tilde{K}$ after the substitution given above. The relations obtained show that for the construction of $p(x,y)$ it is expedient to use as far as possible the uniformly largest of the functions $\{k_{ij}(x,y)\}$.

Let us denote the matrix-integral operator in Eq. (2.6) by the symbol $\tilde{K}_p$. The operator $\tilde{K}_p$ leaves invariant the cone of covariance matrix-functions, and hence it is natural to assume that to the maximal eigenvalue $\rho(\tilde{K}_p)$ there corresponds the eigen-covariance matrix $S(x)$ (for this it is sufficient that the operator $\tilde{K}_p$ should be completely continuous and strongly positive).

Theorem 2.1. *Let the eigen-covariance matrix corresponding to the eigenvalue $\rho(\tilde{K}_p)$ be $S(x)$, where*

$$\|S\| = 1, \quad k_{ij} \ge 0, \quad \text{and} \quad \|\tilde{K}_p\| < 1.$$

Then

$$[1 - \rho(\tilde{K}_p)]^{-1} \le \sup_{\|x\|=1} \|\Psi(x)\| \le [1 - \|\tilde{K}_p\|]^{-1}. \tag{2.7}$$

Proof. The right part of inequality (2.7) is evident, while the left part is true, in particular, if there exists such a H that $\chi = H\Phi' + \Phi H' - HH' = S$, as here the norm of the solution of Eq. (2.6) is equal to $[1 - \rho(\tilde{K}_p)]^{-1}$. Considering the equality $\chi = S$ jointly with Eq. (2.1), we obtain the matrix equation

$$\Phi\Phi' - (\tilde{K}\Phi)(\tilde{K}\Phi)' = S. \tag{2.8}$$

Diagonal elements of Eq. (2.8) from the system

$$\varphi_i^2 = [\tilde{K}\Phi]_i^2 + S_{ii},$$

which is uniquely solvable, as the transformation with the elements $([\tilde{K}\Phi]_i^2 + S_{ii})^{1/2}$ is a contraction in L_∞. Nondiagonal elements of Eq. (2.8) for the function Φ^* thus obtained, generally speaking, do not form a uniquely solvable system. However, the inequalities

$$\varphi_i^*\varphi_j - [\tilde{K}\Phi^*]_i[\tilde{K}\Phi^*]_j \geq S_{ij}, \quad i,j = 1,...,n,$$

turn out to be true. Consequently, if we put $H^* = \Phi^* - \tilde{K}\Phi^*$, then for the corresponding Ψ^* we have

$$\|\Psi\| \geq 1 - \rho(\tilde{K}_p).$$

□

Later, examples will be considered showing that sometimes the bounds in Eq. (2.7) may be sufficiently close for different variants of $p(x,y)$. Therefore, as a practical criterion of uniform optimality of a weighted method (i.e. a Markov chain), it is possible to consider the quantity $\|\tilde{K}_p\|$, the expression for which has been given above, i.e. $p(x,y)$ is appropriately determined by means of the problem

$$\inf_p \sup_i \int \frac{\left(\sum_{j=1}^m |k_{ij}(x,y)|\right)^2}{p(x,y)} dy, \quad x \in X. \tag{2.9}$$

Solution of minimax problems of such a type was be studied in [40].

Consider now the Monte Carlo technique combined with the finite sum method. The integral equation can be written formally as the system

$$\phi(x) = \sum_{j=1}^m \int_{D_j^m} k(x,y)\phi(y)dy + h(x), \quad x \in D_i^{(m)}, \quad i = 1,...,m. \tag{2.10}$$

Hence, Eq. (2.10) takes the form of equation (2.1) by an appropriate change of variables. The asymptotic behaviour of the cost of the corresponding vector algorithms, as the diameter $d(D_i^{(m)})$ of $D_i^{(m)}$ uniformly decreases, was investigated in [40]. These asymptotic estimates show that an optimal value of m exists if the dimension m is greater than 1. It is therefore appropriate to choose some optimal m in the method of finite sums and then perform averaging over independent realizations of the corresponding randomized algorithm. This technique was applied to solve the integral transfer equation for a plane layer $0 \leq z \leq T$ [40].

Vector estimates for triangular matrix kernel are considered below. Consider a system of the form

$$\chi_i = h_i + \sum_{j=1}^i K_{ij}\chi_j, \quad i = 1,...,m, \tag{2.11}$$

where K_{ij} is an integral operator with the kernel $k_{ij}(x,y)$.

Theorem 2.2. *If K_{ij} are bounded operators, then the inequality*

$$\rho(\tilde{K}) \leq \max_i \rho(K_{ii}) = \rho_0$$

holds. Furthermore, if $k_{ij}(x,y) \geq 0$, then $\rho(\tilde{K}) = \rho_0$.

Proof. Let us consider the system

$$\chi_i = h_i + \lambda \sum_{j=1}^{i} K_{ij}\chi_j, \quad i = 1, ..., m.$$

This system is uniquely solvable, if $0 < \lambda < \rho_0$. Therefore, $\rho(\tilde{K}) < \rho_0$ (see, for example, [10]). If $k_{ij}(x,y) \geq 0$, then $\sup \|K_{ii}^n\| < \|\tilde{K}^n\|$ and $\rho_0 \leq \rho(\tilde{K})$. This completes the proof. □

Note that the first statement of Theorem 2.2 holds for arbitrary linear operators. Denote by $K_{p,ij}$ an integral operator with kernel $k_{ij}^2(x,y)/p(x,y)$.

Theorem 2.3. *If $K_{p,ij}$ are bounded operators, then system Eq. (2.11) satisfies the inequality*

$$\rho(\tilde{K}_p) \leq \max_i \rho(K_{p,ii}) = \rho_0^{(p)}.$$

If, in addition, $k_{ij}(x,y) \geq 0$, then $\rho(\tilde{K}_p) = \rho_0^{(p)}$.

Proof is similar to the proof of the Theorem 2.2. □

It follows from Theorem 2.3 that the variances $D\xi_{j,\pi}$ of the estimates of the solution components for system Eq. (2.11) are finite if, for example, $\|K_{p,ii}\| < 1, \quad i = 1, 2, ..., m$. This gives a simple condition of variance finiteness for the vector algorithm for the multigroup transfer equations proposed in [52]. Here, the transition into another group is treated as absorption. Therefore the equality

$$\max_i \|K_{p,ii}\| = \|K_{p,11}\|.$$

2.2. CALCULATION DERIVATIVES AND PERTURBATIONS WITH RESPECT TO PARAMETERS

The kernels of the integral equations often depend on parameters describing certain variable characteristics of the original information, for example, properties of the medium where the process is considered. In this connection, various problems, for example, inverse problems of prediction of the real values of parameters can be considered using the observations of some functionals of the solution [2,28]. Besides, optimization problems with respect to parameters for these functionals are considered. When the parameters are random, we also consider the problem of calculation of probabilistic characteristics of the solution. It is not difficult to see that to solve these problems, it is useful to have an algorithm for calculating perturbations of the solution or the functionals under perturbations of the parameters. Sometimes it is more effective to calculate derivatives of the solution (or functionals) with respect to parameters. Below, we present a special vector technique for calculating the derivatives and the perturbations mentioned.

Consider in $L_\infty(X)$ an integral equation of the second kind with a kernel depending on a parameter λ:

$$\varphi(x,\lambda) = \int_X k(x,y,\lambda)\varphi(y,\lambda)dy + h(x,\lambda). \tag{2.12}$$

Rewrite Eq. (2.12) in the operator form

$$\varphi = K\varphi + h.$$

In spite of the integral operator in Eq. (2.12) having conjugate form, it is here denoted by the symbol K for simplicity.

In Eq. (2.12), X is a n-dimensional Euclidean space. Consider the problem of calculating the derivatives

$$\varphi^{(n)}(x,\lambda) = \frac{\partial^n(x,\lambda)}{\partial\lambda^n}.$$

Introduce the following notation: $K^{(n)}$ is the integral operator with the kernel $k^{(n)}(x,y,\lambda)$, and $\rho(K)$ is the spectral radius of the operator K. It is assumed that the functions $k^{(n)}$ and $h^{(n)}$ are measurable with respect to x. By formally differentiating relation Eq. (2.12) n-times with respect to λ, we obtain a triangular system of integral equations

$$\varphi^{(n)} = \sum_{i=0}^{n} C_n^i K^{(n-i)}\varphi^{(i)} + h^{(n)}, \quad n = 0,1,...,m \tag{2.13}$$

or in the operator form $\Phi = \tilde{K}\Phi + H$.

Now we give a justification of the utilization of equation (2.13) for calculating the derivatives $\varphi^{(n)}$.

Introduce the additional notation: $F_m(f)$ is the column vector consisting of the derivatives of the function $f(\lambda)$ with respect to λ, i.e.

$$F_m(f) = (f, f^{(1)}, ..., f^{(m)})^T$$

, $D_m(f) = d_{ni}$ is the lower triangular $(m+1)$-dimensional matrix with entries

$$d_{ni} = C_n^{n-i} f^{(n-i)}, \quad i = 0,1,...,n; \quad n = 0,...,m.$$

Using the Leibnitz formula for the m-th derivative of the product of two functions, we obtain

Lemma 2.1. *The following relation holds:*

$$F_m\left(\Pi_{i=0}^k f_i\right) = D_m(f_k)\cdots D_m(f_1)F_m(f_0).$$

□

Theorem 2.4. *Suppose that the inequalities*

$$|k^{(n)}(x,y;\lambda')| \le k_n(x,y), \quad |h^{(n)}(x,\lambda)| \le h_0(x), \quad h_0 \in L_\infty \tag{2.14}$$

hold in the interval $\lambda - \varepsilon < \lambda' < \lambda + \varepsilon$ for some $\varepsilon > 0$, the integral operators K_n with the kernels $k_n(x,y)$ are bounded, $n = 0,...,m$ and $\rho(K_0) < 1$. Then $\rho(\tilde{K}) < 1$ and the functions $\varphi^{(n)}$ satisfy system Eq. (2.13), $n = 0,1,...,m$.

Proof. Following from the Theorem 2.2, the relation $\rho(\tilde{K}) \le \rho(K_0) = \rho_0 < 1$ holds in the interval $(\lambda - \varepsilon, \lambda + \varepsilon)$. By Lemma 2.1, the corresponding Neumann series for system Eq. (2.13) coincides with the series obtained by the formal differentiation of the Neumann series for equation (2.12). As follows from Eq. (2.14), this series is uniformly majorized in $(\lambda - \varepsilon, \lambda + \varepsilon)$. The Neumann series for the majorant is finite thus proving Theorem 2.4. □

Lemma 2.1 and Theorem 2.4 can be directly extended to the case of vector parameter $\lambda = (\lambda_1, ..., \lambda_k)$, since the corresponding systems of type Eq. (2.13) also can be written in a triangular form with the diagonal entries coinciding with K. To do this, it is sufficient to choose an indexation of the functions $\varphi^{(i_1,...,i_k)}$ according to the natural order they are written in the equations. This kind of indexation is universal provided that the equations are written in accordance with the natural increasing order of the derivatives (note that inside a fixed order of the derivative the indexation can be arbitrary).

Let us now mention some problems where this technique of caclulating derivatives can be applied. Suppose that σ is a random vector and the functionals $J_k(\sigma)$ which we study depend on $\sigma_i (i = 1, ..., r)$ almost linearly. Note that the last assumption can be considered in the neighbourhood of the point $\sigma^{(0)} = M\sigma$ (the probability of hitting this neighbourhood is close to unity). Then the correlation moments of the vector $J = J_k(\sigma)$ and the autocorrelation moments of the vectors σ and J can be calculated by the linearization of $J(\sigma)$ using the Monte Carlo evaluations of the derivatives $\partial J_k / \partial \sigma_i$. Note that this calculation technique is developed in [2] for the transfer theory problems for the case of the piecewise constant random scattering cross section σ. The calculation of the higher order derivatives thus can be used here to control the numerical results and to make them more accurate. In [2,28] we present also examples of iterative numerical solution of an inverse problem of finding σ by linearizing the equation

$$J_k(\sigma) = J_k^{(0)}, \quad k = 1, ..., n \ge r,$$

where $J_k^{(0)}$ are known (measured) values of functionals. Note that it is, in fact, a numerical realization of the inverse problems formulated in [2,28] (see Section 5.8).

Consider now a possibility of utilizing estimates of the derivatives for solving special nonlinear integral equations.

Suppose that the parameter σ depends on the solution, i.e. $\sigma = S(\varphi)$; σ thus satisfies the following nonlinear equation:

$$\sigma = S[\varphi(\sigma, \cdot)]\,. \tag{2.15}$$

Linearizing $\varphi(\sigma, \cdot)$ in Eq. (2.15) with respect to σ at the point $\sigma = \sigma^{(0)}$, we obtain a simpler equation

$$\sigma = S[\varphi(\sigma^{(0)}, \cdot) + \varphi'_\sigma(\sigma^{(0)}, \cdot)(\sigma - \sigma^{(0)})]\,. \tag{2.16}$$

The solution of this equation gives the first approximation $\sigma = \sigma^{(1)}$. Relation Eq. (2.16) thus determines an iteration procedure for solving equation (2.15). This process may converge far more rapidly than the Newton iteration process for Eq. (2.15) since the procedure described solves equation (2.15) in a partly exact manner. Note that this technique can be used, among other things, for solving nonlinear problems of radiation heat transfer. Besides, calculation of higher order derivatives also can be used to make numerical results more accurate.

Consider now a problem of calculating the derivative with respect to some direction ω, i.e. of the quantity

$$\varphi'(x) = \frac{d\varphi(x+t\omega)}{dt}|_{t=0}\,.$$

To this end, we differentiate with respect to t the relation

$$\varphi(x+t\omega) = \int_X k(x+t\omega, y)\varphi(y)dy + h(x+t\omega)\,. \tag{2.17}$$

Interchanging the order of differentiation and integration, we obtain

$$\varphi'(x) = \int_X k'(x,y)\varphi(y)dy + h'(x)\,, \tag{2.18}$$

where

$$k'(x,y) = \frac{dk(x+t\omega, y)}{dt}|_{t=0}\,.$$

This interchanging must be additionally justified in all particular cases. Using standard Monte Carlo algorithms for solving equation (2.18), it is not difficult to construct probabilistic estimates for the quantity $\varphi'(x)$ (see, for example, Sect. 1.2).

However, in the radiative transfer theory which is an important area of applications of the Monte Carlo methods, this technique fails because the kernel of the integral equation involves a 'δ-factor' which connects the new collision point with the particle direction (see Section 5.1).

Therefore, it is useful to make the change of variables $x + t\omega \to x$; then equation (2.17) is transformed to the following equation:

$$\psi(x,t) = \int_X k_1(x,y,t)\psi(y,t)dt + h_1(x,t)\,, \tag{2.19}$$

where

$$\psi(x,t) = \varphi(x+t\omega), \quad k_1(x,y,t) = k(x+t\omega, y+t\omega), \quad h_1(x,t) = h(x,t\omega).$$

This change of variable is convenient in calculations if the original equation is invariant (at least partly) under the translation along the direction ω. Since the variable t is a parameter of equation (2.19), the construction of the triangular system of integral equations for the derivatives

$$\frac{\partial^n \psi}{\partial t^n}\Big|_{t=0} = \frac{\partial^n \varphi}{\partial t^n}\Big|_{t=0}$$

can be carried out in an obvious manner.

Consider now the Monte Carlo algorithms for solving the constructed triangular systems of integral equations. Theorem 2.3 shows that the variances of these algorithms are bounded if the operators with the kernels $k^{(n)^2}(x,y,\lambda)/p(x,y)$ are bounded and the spectral radius of the operator with the kernel $k^2(x,y,\lambda)/p(x,y)$ is less than unity ($n = 1, ..., m$), i.e. if the standard variance finiteness condition for the scalar Monte Carlo estimates for equation (2.12) is satisfied.

As a special generalization of system Eq. (2.1) consider a family of integral equations

$$\varphi(x,\lambda) = \int_X k(x,y,\lambda)\varphi(y,\lambda)dy + h(x,\lambda)$$

depending on a parameter λ.

In this case, the Monte Carlo estimates are scalar values; however, they depend on λ, i.e.

$$\xi_x = \xi_x(\lambda) = \sum_{n=0}^{N} Q_n(\lambda)h(x,\lambda)\,,$$

where $Q_n(\lambda)$ are scalar weights.

Here, a problem of functional convergence of the Monte Carlo estimates arises now. Obviously, it is directly related to the random function

$$\sqrt{S}\Big\{S^{-1}\sum_{i=1}^{s}\xi_x^{(i)}(\lambda) - \varphi(x,\lambda)\Big\}, \quad \lambda_1 \le \lambda \le \lambda_2\,, \tag{2.20}$$

where $\xi_x^{(i)}(\lambda)$ are independent realizations of the random variable $\xi_x(\lambda)$, $i = 1,...,s$.

According to [20], function of Eq. (2.20) is weakly convergent to a continuous Gaussian random function in the interval $\lambda_1 < \lambda < \lambda_2$ if the following conditions hold:

$$\left|\frac{\partial\xi_x(\lambda)}{\partial\lambda}\right| \le G_x, \quad \mathrm{E}G_x^2 < +\infty, \quad \lambda_1 < \lambda < \lambda_2\,. \tag{2.21}$$

Therefore, to study the functional convergence of the Monte Carlo estimates to $\varphi(x,\lambda)$ in some interval $\lambda_1 < \lambda < \lambda_2$, it is sufficient to consider an auxiliary triangular system of two integral equations of type Eq. (2.13) where $k(x,y,\lambda)$ is replaced with $|k(x,y,\lambda)|$ while $k'(x,y,\lambda)$ is replaced with $|k'(x,y,\lambda)|$. This permits the construction of the majorant mentioned in the condition for weak convergence Eq. (2.21). Analogously to Theorem 2.4 we can prove

Theorem 2.5. *Suppose that the inequalities*

$$|k(x,y,\lambda)| \le k_0(x,y), \; |k'(x,y,\lambda)| \le k_1(x,y)$$

hold in the interval $\lambda_1 < \lambda < \lambda_2$. If K_1, the integral operator with the kernel $k_1^2(x,y)/p(x,y)$, is bounded and the spectral radius of $K_{p,0}$, the operator with the kernel $k_0^2(x,y)/p(x,y)$, is less than unity, then condition of Eq. (2.21) is satisfied. □

Remark. Note that the condition $\rho(K_{p,0}) < 1$ can be replaced with a wider assumption, i.e. the variance of the estimate $\xi_{x,0}$, corresponding to the operator K_0, is finite (see the next Section).

We will obtain now a system of equations describing small perturbations of the solution of integral equations of the second kind under small perturbations of the kernel and the right-hand side. Consider two equations

$$\varphi = K\varphi + h, \quad \varphi_1 = K_1\varphi_1 + h_1\,. \tag{2.22}$$

Subtracting yields

$$\varphi_1 - \varphi = K_1(\varphi_1 - \varphi) + (K_1 - K)\varphi + h_1 - h.$$

If we can neglect the term $(K_1 - K)(\varphi_1 - \varphi)$, then we obtain the following system of equations:

$$\varphi = K\varphi + h, \quad \delta\varphi = K(\delta\varphi) + \delta K\varphi + \delta h.$$

It is not difficult to see that system of Eq. (2.22) gives the principal part of the perturbation of φ if the dependence of the functional characteristics of the problem on the varying parameter is smooth enough. Let us suppose for simplicity that $\delta h \equiv 0$. Denote by $R(K)$ the resolvent of the operator K, i.e. we put $R(K) = (I - K)^{-1}$. From Eq. (2.22) we have

$$\delta\varphi = R(K)\delta K R(K)h.$$

This representation of small perturbations of the first order corresponds to the general perturbation theory (see, for example, [10]) which gives the following obviously verified relation:

$$R(K_1)h = \sum_{n=0}^{\infty} [R(K)\delta K]^n R(K)h = \sum_{n=0}^{\infty} \varphi_n, \tag{2.23}$$

provided $\rho[R(K)\delta K] < 1$. It is not difficult to see that the terms of series of Eq. (2.23) satisfy the following triangular system of integral equations:

$$\varphi_0 = K\varphi_0 + h, \quad \varphi_n = K\varphi_n + \delta K\varphi_{n-1}, \quad n = 1, 2, \dots. \tag{2.24}$$

Below, we shall consider finite systems of type of Eq. (2.23), i.e. $n = 1, \dots, m$.

In the next section we shall describe non-biased vector weight estimates for solving systems of integral equations. Note that this technique fails if the kernels of the operators K and K_1 are mutually singular, for example, if they include 'δ-functions' which define integration over different manifolds whose dimensions are less than that of X. In this case, it is perhaps more appropriate to use the following system instead of equation (2.24):

$$\varphi_0 = K\varphi_0 + h,$$

$$\varphi_n = \sum_{i=1}^{n} (\delta K)^{n-i} K\varphi_i + (\delta K)^n h, \quad n = 1, 2, \dots. \tag{2.25}$$

which is obtained by substituting the last expression for φ_n into equation (2.24), $n = 0, 1, \dots$.

Using Eq. (2.24) and Eq. (2.25), one can sometimes calculate the solution of the integral equation $\varphi = K_1\varphi + h$ by the Monte Carlo method even if $\rho(K_1) = 1$, i.e. when the standard Monte Carlo estimates are not applicable.

To do this, it is necessary to construct an operator K, such that

$$\rho(K) < 1, \quad \rho(R(K)\delta K) < 1, \quad \rho(K^{(0)}) < 1, \tag{2.26}$$

where $K^{(0)}$ is the operator with the kernel $|k(x, y)|$. From Eq. (2.23), we obtain

$$\varphi = \lim_{n\to\infty} \psi_n, \quad \psi_n = \sum_{k=0}^{n} \varphi_k.$$

The successive summation of equations (2.24) gives the following system of integral equations for ψ_n:

$$\psi_0 = K\psi_0 + h, \quad \psi_n = K\psi_n + \delta K\psi_{n-1} + h, \quad n = 1, 2, \dots .$$

To solve finite systems of this kind, one can use vector Monte Carlo algorithms if conditions of Eq. (2.26) are satisfied.

The optimization of the method presented for calculation of the function φ consists in obtaining a relation between the size of the system of equations and the size of a sample, which is such that the mean number of operations needed to obtain the result to within a prescribed accuracy is minimal. Note that this optimization problem can be solved similarly to the optimization of the Monte Carlo algorithms for the numerical realization of resolvent iterative processes [40].

Consider now the construction of vector algorithms for system of Eq. (2.23). Representing the matrix weight factor as a matrix multiplied by the kernel $k(x, y)$, we obtain the following relations for evaluating the matrix weights:

$$Q_n = q_n Q'_n, \quad q_0 = 1, \quad q_n = q_{n-1} k(x_{n-1}, x_n)/p(x_{n-1}, x_n),$$
$$Q'_0 = Q_0 = \{\delta_{ij}\}, \quad Q'_n = Q'_{n-1}\Delta[1, \delta K(x_{n-1}, x_n)/k(x_{n-1}, x_n)],$$

where $\Delta[1, a]$ is a bidiagonal matrix with the unity as the diagonal entries; the entries of the subdiagonal are equal to a. Consequently, only two different entries compose two non-zero diagonals of the matrix Q'_n thus essentially decreasing the cost of the algorithm.

It is not difficult to see that the following representation holds:

$$\varphi_1(x) = \mathrm{E}\eta_x, \quad \eta_x = \sum_{n=1}^{N} q_n h(x_n) \sum_{m=0}^{n-1} \frac{\delta k(x_m, x_{m+1})}{k(x_m, x_{m+1})} \tag{2.27}$$

if $\delta h \equiv 0$.

The expression for η_x can be formally obtained from the relation

$$\delta q_n = q_n \delta \ln q_n.$$

It is interesting to note that this procedure can ensure a satisfactory estimation of the perturbations even if $\delta k/k$ is not small everywhere, provided conditions of Eq. (2.26) hold.

2.3. CALCULATION OF PARAMETRIC DERIVATIVES IN A SPECIAL CASE

For a class of problem (see Section 4.1) the condition $\rho(K_1) < 1$ is not fulfilled for the appropriate integral operators, and the following general assertions will be used to justify the required estimates.

Theorem 2.6. *Let the following conditions be met:*

(A) There is such a function $\chi(x) \geq 0$ and such a value $\varepsilon > 0$ that the Neumann series for equation (2.12), after the change $k \to |k|$, $h \to \chi$, converges uniformly with respect to $x \in X$ and with respect to λ on the interval $(\lambda_0 - \varepsilon, \lambda_0 + \varepsilon)$, and

$$|h^{(m)}(x, \lambda)| \leq \chi(x), \quad x \in X, \quad m = 0, 1, \dots, M$$

on the same interval;

(B) The transition density $p(x, y)$ of the terminating Markov chain $\{x_n\}_{n=0}^{N}$ is distinct from zero on the union of the kernels of system (2.13) and

$$\frac{|k^{(m)}(x,y,\lambda)|}{p(x,y)} \le \chi(x), \quad \lambda_0 - \varepsilon \le \lambda \le \lambda_0 + \varepsilon, \quad m = 1, ...M.$$

Then for $\lambda = \lambda_0$ *the following relations hold:*

$$\mathrm{E}\zeta_x = \Phi(x, \lambda_0) = \{\varphi_m(x, \lambda_0)\},$$

$$\varphi_m = \varphi^{(m)}, \quad \tilde{\varphi}^{(m)} = \mathrm{E}|\xi^{(m)}| \le c_m < +\infty, \quad m = 0, 1, ...M.$$

Proof. Introduce the following notation: $\xi^{(m)}$ is the m-th component of the vector estimate ζ for the chain originating at the point x_n,

$$h_n^{(m)} = h^{(m)}(x_n), \quad q_n^{(m)} = \frac{k^{(m)}(x_{n-1}, x_n, \lambda)}{p(x_{n-1}, x_n)}, \quad Q_n = \prod_{k=0}^{m} q_n^{(0)}, \quad q_0^{0)} = 1.$$

Then $\xi^{(m)}$, according to (2.13), can be represented in the form:

$$\xi_0^{(m)} = \sum_{i=0}^{m} C_m^i q_1^{(m-i)} \xi_1^{(i)} + h_0^{(m)}. \tag{2.28}$$

Assume now that the statement of the theorem is true for the index values $m' = 1, ..., m-1$ (for $m' = 0$ it is obvious) and using (2.28) let us prove that it also holds for $m' = m$. Rewrite (2.28) as follows:

$$\xi_0^{(m)} = q_1^{(0)} \xi_1^{(m)} + \sum_{i=0}^{m-1} C_m^i q_1^{(m-i)} \xi_1^{(i)} + h_0^{(m)}.$$

Substituting here the similar expression for $\xi_1^{(m)}$, we obtain

$$\xi_0^{(m)} = q_1^{(0)} q_2^{(0)} \xi_2^{(m)} + \sum_{i=0}^{m-1} C_m^i (q_1^{(0)} q_1^{(m-i)} \xi_2^{(i)} + q_1^{(m-i)} \xi_1^{(i)}) + h_0^{(m)} + q_1^{(0)} h_1^{(m)}.$$

Continuing the recursion, we obtain

$$\xi_0^{(m)} = \sum_{i=0}^{m-1} C_m^i \sum_{n=0}^{N} Q_{n-1} q_n^{(m-i)} \xi_n^{(i)} + \sum_{i=0}^{N} Q_n h_n^{(m)}. \tag{2.29}$$

By virtue of the indication assumption made, we have

$$\mathrm{E}(|\xi_n^{(i)}||x_0, x_1, ..., x_n) = \tilde{\varphi}^{(i)}(x_n) < c_i < +\infty, \quad i = 0, 1, ..., m-1.$$

Using suppositions (A) and (B) and the theory of scalar 'collision' estimate [40], we obtain

$$\mathrm{E}(|\zeta_0^{(m)}|) \le \mathrm{E}\left[\sum_{i=0}^{m-1} C_m^i \sum_{n=1}^{N} |Q_{n-1}| \chi_{n-1} |\tilde{\varphi}_n^{(i)}| + \sum_{n=0}^{N} |Q_n| \chi_n\right] \le C_m < +\infty$$

$$\forall x_0, \quad \lambda \in (\lambda_0 - \varepsilon, \lambda_0 + \varepsilon).$$

Since the expression $\xi_0^{(m)}$ can be obtained by formal differentiation of the expression $\xi_0^{(0)}$ (see [37]), Theorem 2.6 may be considered to be proved. □

Theorem 2.7. *If conditions (A) and (B) of Theorem 2.6 are fulfilled and the variance obeys*

$$\mathrm{D}\left[\sum_{n=0}^{\infty} |Q_n|\chi(x_n)\right] \leq C < +\infty \quad \forall x \in X$$

then

$$\mathrm{D}\xi_0^{(m)} < +\infty, \quad m = 0, 1, 2, \ldots . \tag{2.30}$$

Proof. The statement of the theorem is evident for $m' = 0$. Assume that the statement is also true, for the index values $m' = 1, ..., m-1$. Clearly, to justify the validity of inequality (2.30) it suffices to prove the finiteness of the mean squares of the terms in the first sum in (2.29). These terms can be represented as

$$\begin{aligned}\xi_{0,i}^{(m)} &= \left[\sum_{n=1}^{N} Q_{n-1} q^{(m-i)} \xi_n^{(i)}\right]^2 \\ &= \sum_{n=1}^{N} Q_{n-1}^2 (q_n^{(m-i)})^2 (\xi_n^{(i)})^2 + 2\sum_{n=1}^{N} Q_{n-1}^2 q_n^{(m-i)} \xi_n^{(i)} \sum_{k=n+1}^{N} [Q_{k-1}/Q_{n-1}] q_k^{(m-i)} \xi^{(i)} .\end{aligned}$$

Hence

$$\begin{aligned}\mathrm{E}\xi_{0,i}^{(m)} &= \mathrm{E}\sum_{n=1}^{N} Q_{n-1}^2 (q_n^{(m-i)})^2 \mathrm{E}\{(\xi_n^{(i)})^2 | x_0, ..., x_n\} \\ &+2\mathrm{E}\sum_{n=1}^{N} Q_{n-1}^2 q_n^{(m-i)} \mathrm{E}\left\{\xi_n^{(i)} \sum_{k=n+1}^{N} [Q_{k-1}/Q_{n-1}] q_k^{(m-i)} \xi_k^{(i)} | x_0, ..., x_n\right\}, \\ & i = 1, ..., m-1.\end{aligned}$$

Let the symbol E_n denote the conditional expectation for a fixed value of the vector $x_0, ..., x_n$.

With allowance made for the inequality

$$|\mathrm{E}(\xi\eta)| \leq \sigma\xi\sigma\eta + |\mathrm{E}\xi\mathrm{E}\eta|$$

this expectation is transformed and estimated as follows:

$$\begin{aligned}&\left|\mathrm{E}_n\left\{\sum_{k=n+1}^{N} \xi_n^{(i)} [Q_{k-1}/Q_{n-1}] q_k^{(m-i)} \xi_k^{(i)}\right\}\right| \\ &= \left|\mathrm{E}_n\left\{\sum_{k=n+1}^{N} \mathrm{E}\{[Q_{k-1}/Q_{n-1}] q_k^{(m-i)} \xi_n^{(i)} \xi_k^{(i)} | x_{n+1}, ..., x_k\}\right\}\right| \\ &\leq C_1' \mathrm{E}_n\left\{|\xi_n^{(i)}| \sum_{k=n+1}^{N} |Q_{k-1}/Q_{n-1}|\chi_k\right\} + C_0'\chi_n \leq C'\chi_n.\end{aligned}$$

Using this inequality, the induction hypothesis, and the condition of the theorem, we obtain

$$\mathrm{E}\xi_{0,i}^{(m)} \le C_2'\left\{\mathrm{E}\sum_{n=1}^{N} Q_{n+1}^2\chi_n^2 + 2\mathrm{E}\sum_{n=1}^{N} Q_{n-1}^2\chi_n \sum_{k=n+1}^{N} |Q_{k-1}/Q_{n-1}|\chi_k\right\} < +\infty .$$

Theorem 2.7 is proved. □

Now let us apply the technique under consideration to the estimation of the iterates of the resolvent. To this end, in $L_\infty(X)$ consider equation (2.12) of the form:

$$\varphi(x,\lambda) = \lambda^{-1}\int_X k(x,y)\varphi(x,\lambda)dx + \lambda^{-1}h(x)$$

for $|\lambda| > \rho(K)$. We have

$$\varphi_\lambda = [\lambda - K]^{-1}h$$

and

$$\varphi_\lambda^{(m)} = \frac{d^m\varphi_\lambda}{d\lambda^m} = (-1)^m m!|\lambda - K|^{-(m+1)}h.$$

On the other hand,

$$\varphi_\lambda^{(m)} = \mathrm{E}\zeta_x^{(m)}, \quad \zeta_x^{(m)} = \sum_{n=0}^{N}(-1)^m\frac{(n+m)!}{n!}\lambda^{-(n+1+m)}Q_n h(x_n).$$

The unbiasedness of the estimate $\zeta_x^{(m)}$ and the relationship $\mathrm{D}\zeta_x^{(m)} < +\infty$ are evidently ensured by Theorems 2.6 and 2.7, though the results of [37] would suffice in this case. Finally, we get

$$[\lambda - K]^{-(m+1)}h = (-1)^m\varphi_\lambda^{(m)}/m! = \mathrm{E}\xi_x^{(m)}$$

$$\xi_x^{(m)} = \sum_{n=0}^{N} C_{n+m}^m Q_n h(x_n)\lambda^{-(n+1+m)} .$$

Under rather general conditions (see, for example, [10]) we have

$$\frac{([\lambda - K]^{-(m+1)}h, f)}{([\lambda - K]^{-m}h, f)} \to \frac{1}{\lambda - \lambda_*},$$

where λ^* is the eigenvalue of the operator K, closest to λ. If $f(x)$ is the probability density, then

$$([\lambda - K]^{-(m+1)}h, f) = \mathrm{E}\xi_x^{(m)}$$

and x is a random point distributed according to the density f.

Note that the corresponding Monte Carlo algorithm for estimating the eigenvalues has been suggested in [37], where $1/\lambda$ was considered instead of λ and the question of whether the variance is finite has not been discussed.

Chapter 3

Solution of the Helmholtz equation

3.1. THE 'WALK ON SPHERES' PROCESS

Let us consider a three-dimensional Dirichlet problem for a variable-parameter Helmholtz' equation

$$\Delta u + cu = -g, \quad u\big|_{\Gamma} = \psi \tag{3.1}$$

as a main problem in a domain $D \subset X_3$ with the boundary Γ , where $c(r) < c^*$ and $-c^*$ is the first eigenvalue of the Laplacian operator for the domain D, $r = (x, y, z) \in D$.

The conditions for the functions c, g, ψ, and Γ to be regular are assumed to be fulfilled. These conditions guarantee the existence and uniqueness of the solution of problem (3.1), as well as its probabilistic representation and an integral representation with making use of the Green's function for a ball (see, e.g. [13,18]).

Both new estimates and available ones to be considered below are associated with the so-called walk on spheres process within the domain D [43]. To describe the process, let us introduce the following notation: $\bar{D}$ is a closure of the domain D; $d(P)$ is a distance from the point P to the boundary Γ; Γ_ε is an ε-neighbourhood of the boundary Γ, i.e.

$$\Gamma_\varepsilon = \{P \in \bar{D} : d(P) < \varepsilon\}, \tag{3.2}$$

$S(P)$ is the largest sphere of those centered at the point P and entirely lying in $\bar{D}$, i.e.

$$S(P) = \{Q \in \bar{D} : |Q - P| = d(P)\} .$$

In the walk on spheres process, every next point P_{k+1} is chosen uniformly over the surface of the sphere $S(P_k)$; the process terminates when a point finds itself inside Γ_ε.

Let us denote by $s_0(P, \varepsilon)$ the surface of the sphere $S(P)$ that belongs to the set Γ_ε. Construct a sphere S_ε with radius ε centered at the point of contact of the boundary Γ with the sphere $S(P)$. Then the area of the part of the sphere $S(P)$ that entirely lies inside $S_\varepsilon(P)$ equals $\pi\varepsilon^2$. This yields the following lower bound for the probability of the next point up inside Γ_ε:

$$\frac{s_0(P, \varepsilon)}{4\pi d^2(P)} \geq \frac{\pi\varepsilon^2}{4\pi d^2(P)} \geq \frac{\varepsilon^2}{4d_{\max}^2} = \nu(\varepsilon), \tag{3.3}$$

where $d_{\max}$ is the exact upper boundary of radii of the spheres entirely lying in D.

Now we give a rigorous definition of the walk on spheres process. Let us define a Markov chain $\{P_n\}_{n=1,\dots,N}$ by the following characteristics: $\pi(r) = \delta(r - P_0)$ is the initial distribution density (i.e. the chain originates at the point P_0); $p(r,r') = \delta_r(r')$ is the density of the transition from r to r' which is a generalized three-dimensional density of the uniform probability distribution on the sphere $S(r)$; $p_0(r)$ is the probability of the chain termination; it is determined by the expression:

$$p_0(r) = \begin{cases} 0, & r \notin \Gamma_\varepsilon \\ 1, & r \in \Gamma_\varepsilon , \end{cases}$$

where N is the number of the last state.

This chain described above is called a walk on spheres process. Evidently, it can be represented as follows:

$$P_n = P_{n-1} + \omega_n d(P_{n-1}), \quad n = 1, 2, \dots ,$$

where ω_n is the sequence of independent isotropical unit vectors. The probability $p_1(r)$ of the chain terminating after the transition from r to r' can be estimated as follows:

$$p_1(r) = \int_{\Gamma_\varepsilon} p(r,r')p_0(r')dr' \geq \nu(\varepsilon). \tag{3.4}$$

Hence, the mean number $\mathrm{E}N = q(P_0, \varepsilon)$ of the transitions in the walk on spheres chain, which governs the average time it takes for a computer to model the chain, does not exceed $\nu^{-1}(\varepsilon)$.

Moreover, for a wide class of boundaries Γ, the following logarithmic estimate has been obtained on the basis of the restoration theory [13, 18]:

$$q(P_0, \varepsilon) \leq C|\ln \varepsilon| \ . \tag{3.5}$$

This estimate is obviously valid for convex domains. Note that for certain non-convex domains with corners, the estimates of the form of (3.5) have been obtained in [34].

Estimate (3.5) becomes heuristically evident from the following consideration of the spheres' centres. Muller [43] has shown, that the walk on spheres process, with a probability close to unite, converges to the boundary Γ, therefore, it is sufficient to consider the density near a plane boundary.

Let x be a distance to the plane boundary. From similarity considerations it follows that the density $f(x)$ of the distribution of spheres' centres in x should be close to x^{-1} up to a constant factor.

Simple reasoning yields that the density $f(x)$ of the distribution of spheres' centres in the distance x from the plane boundary, when $x \to 0$, approaches the positive solution to an integral homogeneous equation with a kernel equal to the corresponding transition density. Let us write down this equation for two- and three-dimensional cases, respectively:

$$f^{(2)}(x) = \frac{1}{2\pi} \int_{x/2}^{\infty} \frac{f^{(2)}(x')dx'}{\sqrt{[x(2x'-x)]}}, \quad f^{(3)}(x) = \int_{x/2}^{\infty} \frac{f^{(3)}(x')dx'}{2x'} \ .$$

To derive these equations it is sufficient to verify that the density of the transition from x' to x equals the integrand. Direct substitution shows that $f(x) = Cx^{-1}$ satisfies both equations.

The conclusion drawn has been verified by direct summation of the densities $f_m^{(3)}(x)$ of the distributions of the centres of the spheres with the multiplicity of m for a given 'source' located at a point $x_0 > 0$. The following expression can be easily obtained:

$$f_m^{(3)}(x) = \frac{1}{2^{m+1}x_0}\int_{x/2}^{2^m x_0}\cdots\int_{x_{m-1}/2}^{2x_0}\frac{dx_1\ldots dx_m}{x_1x_2\ldots x_m} = \frac{[\ln(2^{m+1}x_0/x)]^m}{2^{m+1}m!x_0}, \quad x < 2^{m+1}x_0 .$$

Hence,

$$f_m^{(3)}(x) = \frac{1}{2x_0}\sum_{m=0}^{\infty}\frac{[\ln(2^{m+1}x_0/x)]^m}{2^m m!} . \tag{3.6}$$

The latter series converges for any $x > 0$. The sum of the first 100 terms of the series has been calculated with $x_0 = 1$: for $x = 0.01, 0.02, \ldots, 0.09$ we obtained the values of $f^{(3)}(x) = 324.6$, 162.4, 108.3, 81.28, 65.04, 54.21, 46.47, 40.66, and 36.15, respectively.

The values of $f^{(3)}(x)$ obtained to three decimal places obey the law $3.253 \cdot x^{-1}$. Hence it follows that

$$\int_{\varepsilon}^{a} f(x)dx \sim C|\ln\varepsilon|, \quad a > 0.$$

Note that the above reasoning, however, does not prove but does explain estimate (3.5). Below, in Section (3.7), similar considerations will be used for additional justification of the possibility of constructing unbiased estimates on the walk on spheres process.

Muller [43] has also justified the use of the walk on spheres process for solving the Laplace equation. On the basis of the probabilistic representation of the solution of Dirichlet problem [25], Elepov and Mikhailov [14] used this process for solving Helmholtz' equation with $c = \text{const} < 0$, while Elepov *et al.* [13] extended it to $c = \text{const} < c^*$. In [15], with $c = \text{const} \leq 0$ the corresponding algorithms have been constructed and further developed on the basis of the integral representation connected with the Green's function for a ball. Section 3.2 and 3.3 deal with further development of these approaches and contain new results concerning the possibilities offered by the estimates constructed previously.

Ermakov *et al.* [18] used integral representations for the case of the variable parameter $c(r) < 0$. In the principal section of this Chapter (Section 3.4) on the basis of a special integral representation and results of Section 3.2, efficient Monte Carlo algorithms are constructed and justified for the case of the non-negative parameter $c(r)$. The algorithms can be used for solving equations of the higher order and some non-stationary problems (see Section 3.7).

3.2. THE USE OF PROBABILISTIC REPRESENTATION

For $c = \text{const} < c^*$, the probabilistic representation of the solution of problem (3.1) has the form [25]:

$$u(r_0) = \mathrm{E}\int_0^{\tau} e^{ct}g(\xi(t))dt + \mathrm{E}[e^{c\tau}\psi(\xi(\tau))],$$

where $\xi(t)$ is the diffusion process originating at the point r_0 and corresponding to the Laplacian operator, τ is the instant the process leaves the domain D for the first time. Proceeding from the process being strictly Markovian, we have:

$$u(r_0) = \mathrm{E}\sum_{i=0}^{\infty}\int_{\tau_i}^{\tau_{i+1}} e^{ct}g(\xi(t))dt + \mathrm{E}\left[\psi(\xi(\tau))\prod_{i=0}^{\infty} e^{c(\tau_{i+1}-\tau_i)}\right]$$

$$= \sum_{i=0}^{\infty}\mathrm{E}\left[e^{c\tau_i}\int_{0}^{\tau_{i+1}-\tau_i} e^{ct}g(t+\tau_i)dt\right] + \mathrm{E}\left[\psi(\xi(\tau))\prod_{i=0}^{\infty} e^{c(\tau_{i+1}-\tau_i)}\right],$$

where τ_i is the instant the process $\xi(t)$ for the first time arrives at the surface of the i-th sphere of the corresponding walk on the spheres $\{r_n\}$, $n = 0, 1, \ldots, N$, and $r_N \in \Gamma_\varepsilon$.

Let us denote by P the point of the boundary Γ closest to r_N. Using the repeated averaging procedure and the integral representation of problem (3.1) at the ball centre, we can easily obtain [13,14] that

$$u(r_0) \approx u_\varepsilon(r_0) = \mathrm{E}\eta_\varepsilon^{(0)},$$

where

$$\eta_\varepsilon^{(0)} = \sum_{i=0}^{N}\left[\prod_{j=0}^{i-1} s(c,d_j)\right]\int_{D(r_i)} G(\rho;c,d_i)g(\rho)d\rho + \left[\prod_{j=0}^{N-1} s(c,d_j)\right]\psi(P). \tag{3.7}$$

Here $d_j = d(r_j)$, $D(r_i)$ is the ball of radius d_i centered at the point r_i, $G(\rho;c,d)$ is the Green function for a ball and

$$s(c,d) = \begin{cases} d\sqrt{c}/\sin(d\sqrt{c}), & c \geq 0, \\ d\sqrt{|c|}/\mathrm{sh}(d\sqrt{|c|}), & c \leq 0. \end{cases}$$

The validity of the repeated averaging procedure becomes evident if it is considered that the quantities $\tau_{i+1} - \tau_i$ are arbitrarily (at fixed $\{r_i\}$) independent and that the distribution of a quantity $\tau_{i+1} - \tau_i$ does not depend on the position of the point $\xi(\tau_{i+1})$ on the sphere $S(r_i)$. If the first derivatives of the solution are bounded in $\bar{D}$, then

$$|u(r) - u_\varepsilon(r)| \leq C\varepsilon, \quad r \in D.$$

Futher, each of the integrals appearing in (3.7) can be estimated by its own random node ρ_i [15]. It is reasonable to choose random nodes in $D(r_i)$ with the density proportional to $G(\rho;c,d_i)$. For $c \leq 0$ this density will be considered in Section 3.3. In this case, instead of $\eta_\varepsilon^{(0)}$ we obtain the estimate

$$\eta_\varepsilon = \sum_{i=0}^{N}\left[\prod_{j=0}^{i-1} s(c,d_j)\right]g(\rho_i)F_{r_i} + \left[\prod_{j=0}^{N-1} s(c,d_j)\right]\psi(P),$$

where

$$F_r = \int_{D(r)} G(\rho;c,d(r))d\rho \leq C_1 d^2(r). \tag{3.8}$$

The repeated averaging shows that

$$\mathrm{E}\eta_\varepsilon = \mathrm{E}\eta_\varepsilon^{(0)} = u_\varepsilon(r_0).$$

Now let us consider the question of uniform in ε boundedness of the variance $D\eta_\varepsilon$. The variance determines the mean-square error of the resultant Monte Carlo estimate obtained by averaging the realizations of η_ε. This question will be considered here for $c > 0$, as it is quite easy to solve for $c \leq 0$ with the integral representation method (see Section 3.3).

First, assume that $g \equiv 0$.

In this case

$$\eta_\varepsilon \leq \eta = C_2\left[\prod_{j=0}^{\infty} s(c, d_j)\right]$$

and $E\eta < +\infty$ for $c < c^*$ by virtue of the probabilistic representation of the solution of problem (3.1), by analogy with the transition from this representation to $\eta_\varepsilon^{(0)}$.

Hence,

$$E\eta_\varepsilon^2 \leq C_2^2 E\left[\prod_{j=0}^{\infty} s^2(c, d_j)\right].$$

Now let us show that for $c > 0$ the following inequality holds

$$s^2(c/2, d) \leq s(c, d) \tag{3.9}$$

with $d\sqrt{c} < \pi$. The latter requirement is evidently met, if $c < c^*$. Indeed, the quantity c_* decreases with expanding the domain, and for a ball D^* of radius $d_{\max}$ the relationship $d_{\max}\sqrt{c^*(D^*)} = \pi$ is valid.

On replacing $d\sqrt{c} = x$, by substituting the expression $s(c, d)$ into (3.9), we obtain the following inequality:

$$\varphi(x) = x \sin x - 2\sin^2(x/\sqrt{2}) \leq 0. \tag{3.10}$$

Since

$$\frac{d\sin^2(x/\sqrt{2})}{dx} = \frac{1}{\sqrt{2}}\sin(x\sqrt{2})$$

the expansion of the function $\sin^2(x/\sqrt{2})$ into a Maclaurin series is obtained by integrating the corresponding series for the function $\sin(x\sqrt{2})/\sqrt{2}$. As a result, we obtain

$$\begin{aligned}\varphi(x) &= \sum_{k=1}^{\infty}(-1)^{k+1}\left[\frac{1}{(2k-1)!} - \frac{2^k}{(2k)!}\right]x^{2k} \\ &= \sum_{k=1}^{\infty}(-1)^{k+1}\frac{k2^{1-k}-1}{(2k-1)!!k!}x^{2k} = -\frac{x^6}{360} + \dots .\end{aligned}$$

A monotone decrease of the absolute values of the members in this series can be verified in a simple way at $x = \pi$; the first seven absolute values are: 2.67, 1.88, 0.568, 0.100, 0.0119, 0.00103, and 0.0000686.

Hence, $\varphi(x) \leq 0$ for $0 \leq x \leq \pi$, i.e. inequality (3.9) holds, and the uniform in ε boundedness of the variance $D\eta_\varepsilon$ has been proved for $c < c^*/2$ at $g \equiv 0$. Note that this result (see [47]) has been obtained previously, however, in a much more intricate way. The result is obvious because here η is the conditional mean of $\xi = e^{c\tau}\psi(\xi(\tau))$ and $D\xi < +\infty$ if $c < c^*/2$. But the above reasoning clarifies the structure of the quantity $E\eta_\varepsilon^2$ and demonstrates that the condition with $c < c^*/2$ does not seem to be necessary here, i.e. the variance of the estimate may be finite even for the values of c, somewhat over $c^*/2$, especially for the points r_0 close to the boundary.

On the other hand, the quantity $D\eta_\varepsilon$ should as a rule be unbounded for $c > c_0/2$, where c_0 is a solution to the equation:

$$s^2(c_0/2, d_{\max}) = s(c^*, d_{\max})$$

since in this case $c_0 > c^*$, and

$$s^2(c,d) \geq s(c^*, d)$$

(see Section 6.2). Moreover, here $D\eta_\varepsilon = +\infty$ for sufficiently small ε due to the above correspondence between the walk on spheres estimates, the diffusion-process estimates and the relation:

$$c^*(D - \Gamma_\varepsilon) \to c^*(D) = c^*, \quad \varepsilon \to 0.$$

Now let us turn to problem (3.1) for an inhomogeneous Helmholtz equation. Evidently, it will suffice to prove uniform boundedness of the variance for $g \equiv \text{const}$, and the statement is true if, for $\varepsilon = 0$, the variance of the quantity

$$\eta_n = \sum_{i=0}^{n} \left[\prod_{j=0}^{i-1} s(c, d_j)\right] d_i^2$$

is bounded uniformly in n.

Let us consider the relationship:

$$\prod_{j=0}^{n} s(c,d_j) = \prod_{j=0}^{n-1} s(c,d_j) + [s(c,d_n) - 1] \prod_{j=0}^{n-1} s(c,d_j)$$

$$= \ldots = 1 + \sum_{i=0}^{n} \left[\prod_{j=0}^{i-1} s(c,d_j)\right] [s(c,d_j) - 1] = \zeta_n .$$

As $d_{\max}\sqrt{c} < \pi$, then

$$s(c,d_i) - 1 \geq C_3 d_i^2, \quad \eta_n \leq C_3^{-1} \zeta_n .$$

From the above we have

$$\lim_{n\to\infty} D\zeta_n < +\infty$$

and the quantity $D\eta_n$ possesses the same property.

Let us dwell on the results of methodical calculations.

The Monte Carlo method was used to solve the Dirichlet problem in a cube $0 \leq x, y, z \leq 1$ for the equation:

$$\Delta u + cu = 0, \quad u\big|_\Gamma = \cos\left(x\sqrt{\frac{c}{3}}\right) \cos\left(y\sqrt{\frac{c}{3}}\right) \cos\left(z\sqrt{\frac{c}{3}}\right)\Big|_\Gamma . \tag{3.11}$$

The exact solution has the form:

$$u(x,y,z;c) = \cos\left(x\sqrt{\frac{c}{3}}\right)\cos\left(y\sqrt{\frac{c}{3}}\right)\cos\left(z\sqrt{\frac{c}{3}}\right).$$

The solution was estimated at a point r with the coordinates $x = y = z = 0.9$ for different values of c; in so doing for $c \leq 30$ the value $\varepsilon = 0.0001$ was used, while for $c \geq 35$ the value $\varepsilon = 0.01$ was used. It is known that for a unite cube $c^* = 3\pi^2 \approx 29.609$. The results obtained are included into Table 3.1, where N is the number of realizations of walk on spheres, y_N is the estimate of the solution, σ_N is the estimate of the corresponding mean-square error and σ^2 is the estimate of the quantity $D\eta_\varepsilon$.

Table 3.1. Results of calculations for problem (3.11)

c	$-u(r;c)$	N	$-y_N \pm \sigma_N$	σ^2	$N = 4000$	
					$-y_N \pm \sigma_N$	σ^2
35	0.993	16.7×10^6	1.47 ± 0.57	5.4×10^6	0.991 ± 0.075	881
30	0.875	9.8×10^6	0.95 ± 0.05	2.4×10^3	0.876 ± 0.011	4.8
25	0.627	2.5×10^6	0.628 ± 0.006	78	0.627 ± 0.003	0.44
20	0.3197	1.0×10^6	0.3187 ± 0.0010	1.05	0.320 ± 0.0017	0.11
15	0.0781	4.0×10^5	0.0779 ± 0.0005	0.084	0.0785 ± 0.0008	0.023
10	0.00038	2.4×10^5	0.00051 ± 0.00017	0.0072	0.00098 ± 0.00038	0.0057

It is interesting to note that insignificant deviations (with allowance made for statistical estimate σ_N) from the exact solution have been obtained for $c > c_0/2 \approx 18.56$ as well. Insignificant deviations also occur for a smaller sample size $N = 40000$, though the estimate of the quantity σ^2 demonstrates an evidently unstable behavior for $c \geq 15$. The analysis of the results shows that the variance of the estimate for $c > c^*$ diverges in this case but slightly and therefore to 'cut off' the quantity η_ε is worthwhile

$$\eta_\varepsilon \to M\,\mathrm{sign}(\eta_\varepsilon), \quad \text{if} \quad |\eta_\varepsilon| \geq M.$$

The result obtained for problem (3.11) with $M = 100$ and $N = 40000$ are listed in Table 3.2.

Table 3.2. Results of calculations for problem (3.11) with $M = 100$ $(N = 40000)$

c	$-u(r;c)$	$y_N \pm \sigma_N$	σ^2
45	0.834	0.810 ± 0.021	18
40	0.969	0.950 ± 0.017	12
35	0.993	0.976 ± 0.016	7

In the calculations we assumed: $d_i \to 3/\sqrt{c}$, if $d_i\sqrt{c} > 3$.

Hence, the suitable 'cut off' may serve as an efficient regularizer of the algorithm being considered. The problem was also solved in the unit cube:

$$\Delta u + cu = 0, \quad u\Big|_{\Gamma_1} \equiv 1, \quad u\Big|_{\Gamma_i} \equiv 0, \quad i = 2, \ldots, 6, \tag{3.12}$$

where Γ_i are the cube faces, and $\Gamma_1 : x \equiv 1$. The solution was estimated at the point $x = y = z = 0.5$ for $\varepsilon = 0.001$ (see Table 3.3).

Table 3.3. Results of solving problem (3.12)

c	N	$y_N \pm \sigma_N$	σ^2
19	1×10^6	0.726 ± 0.006	32
17	7×10^5	0.583 ± 0.004	9.0
16	1×10^6	0.528 ± 0.002	4.7
14	7×10^5	0.438 ± 0.002	1.92

There exists no exact solution of this problem, but the results agree satisfactorily with the approximate values given in Section 3.3 that have been obtained with making use of the expansion of $u(r;c)$ into a Taylor series at the point $c = c_0 = 14$. It is noteworthy that in this case the estimate of the quantity σ^2 reveals a stable behaviour for the values $c = 16, 17$ as well.

3.3. THE USE OF INTEGRAL REPRESENTATIONS

Let us consider the following tree-dimensional Dirichlet problem:

$$\Delta u - cu = -g, \quad u\big|_\Gamma \equiv \psi \tag{3.13}$$

in a domain D with the boundary Γ $c = \text{const} \geq 0$. The conditions for the functions g, ψ, and Γ to be regular are assumed to be satisfied. These conditions guarantee the existence and uniqueness of the solution of problem (3.13), as well as the possibility of its representation with making use of the Green function for a ball. As previously, we denote

$$\Gamma_\varepsilon = \{r \in D \cup \Gamma : d(r) < \varepsilon\} .$$

The following integral equation can be written for the function $u(r)$ [15]:

$$u(r) = \int_D k(r, r')u(r')dr' + h(r), \tag{3.14}$$

where

$$k(r,r') = \begin{cases} \frac{d\sqrt{c}}{\mathrm{sh}(d\sqrt{c})}\delta_r(r'), & r \notin \Gamma_\varepsilon \\ 0, & r \in \Gamma_\varepsilon \,, \end{cases}$$

$$h(r) = \begin{cases} \frac{1}{4\pi}\int \frac{\mathrm{sh}[(d-|r'-r|)\sqrt{c}]}{|r'-r|\,\mathrm{sh}(d\sqrt{c})} g(r')dr', & r \notin \Gamma_\varepsilon \\ u(r), & r \in \Gamma_\varepsilon \,. \end{cases}$$

Here $d = d(r)$ and $\delta_r(r')$ is the generalized density corresponding to the uniform distribution of probabilities over the sphere $S(r)$. As $d\sqrt{c}/\,\mathrm{sh}(d\sqrt{c}) \leq 1$ for $c \geq 0$, then

$$\begin{aligned} \int_D\int_D k(r,r')k(r',r'')dr''dr' &\leq \int_{D-\Gamma_\varepsilon} \delta_r(r')\Big(\int_D \delta_{r'}(r'')dr''\Big)dr' \\ &= \int_{D-\Gamma_\varepsilon} \delta_r(r')dr' \leq 1 - \nu(\varepsilon) \end{aligned}$$

by virtue of condition (3.3). Hence, in the space L_∞, which is natural here, we have

$$\|K^2\| \le 1 - \nu(\varepsilon) < 1.$$

This ensures the convergence of the Neumann series for equation (3.14) and thereby makes it possible to use the Monte Carlo method. Besides, it follows here from that the solution of the initial differential problem is a unique solution to equation (3.14). In this case, generalized solutions are also taken into consideration. Therefore, the following relationship can be applied in the estimation:

$$u(r_0) = \mathrm{E}\zeta, \quad \zeta = h(r_0) + \sum_{n=1}^{N} Q_n h(r_n).$$

Here $\{r_n\}$ is the chain of isotropic walk on spheres that terminates inside Γ_ε. For $\varepsilon > 0$ it is reasonable to speak of the chain as an ε-spherical process, with the weights being determined by the formulae:

$$Q_0 = 1, \quad Q_n = Q_{n-1} \frac{d_{n-1}\sqrt{c}}{\mathrm{sh}(d_{n-1}\sqrt{c})} \le Q_{n-1}\,,$$
$$d_n = d(P_n), \quad n = 1, 2, \dots\,.$$

Now we may recall that we have made for a while a non-realistic assumption: the solution $u(r)$ is known in Γ_ε. However, instead of exact values of $u(r)$ in Γ_ε we may use approximate values, e.g. those taken at the nearest points on the boundary, i.e. we may assume

$$u(r) \approx \psi(r^*), \quad r \in \Gamma_\varepsilon, \quad r^* \in \Gamma, \quad |r - r^*| = d(r).$$

As a result, we obtain the biased estimate ζ_ε whose mean value $u_\varepsilon(r_0)$ differ from $u(r_0)$ by the value of the order of ε. Indeed, if we consider the expression for the difference of the estimates ζ and ζ_ε, we obtain

$$|u - u_\varepsilon| \le |\mathrm{E}\{Q_N[u(r_N) - \psi(r_N^*)]\}| \le A\varepsilon,$$

where A is a constant, which is finite due to the solution derivatives being bounded in the domain D (the relation $Q_N \le 1$ has been used here).

In this case it is quite simple to show that the quantity $\mathrm{D}\,\zeta_\varepsilon$ is uniformly bounded. Indeed, the quantity ζ_ε monotonically decreases with $c \to 0$, while at $c = 0$ we have $Q_N \equiv 1$ and (see, e.g. [40])

$$\mathrm{E}\,\zeta_\varepsilon^2 = h_\varepsilon(2u_\varepsilon - h_\varepsilon) + K_\varepsilon(\mathrm{E}\zeta_\varepsilon^2).$$

Hence,

$$\mathrm{E}\zeta_\varepsilon^2 \le K_\varepsilon(\mathrm{E}\zeta_\varepsilon^2) + C|h_\varepsilon|$$

and thereby the quantity $\mathrm{E}\zeta_\varepsilon^2$ can be majorized by the solution to the equation:

$$\chi = K_\varepsilon \chi + C h_\varepsilon^{(1)},$$

which corresponds to the initial problem with the replacement:

$$g \to C_0|g|, \quad \psi \to C_0|\psi| \,.$$

Thus, the quantity $D\zeta_\varepsilon$ is uniformly bounded here, if after the above replacement the generalized solution remains bounded, which is obviously true for $c \geq 0$ and with the above-mentioned regularity of the problem statement.

Futher, the integral determining $h(r)$ for $r \notin \Gamma_\varepsilon$ can be estimated using the Monte Carlo method by a single randomly chosen node. It is natural to sample the node using a polar coordinate system ρ, ω with the density proportional to the Green function (see the preceding section). We denoted by ρ the distance from the next sphere centre (ω is an isotropic unit vector).

Hence, instead of ζ_ε we obtain the random variable

$$\zeta_{\varepsilon,1} = h_1(r_0, \rho_0, \omega) + \sum_{n=1}^{N-1} Q_n h_1(r_n, \rho_n, \omega_n) + Q_N \psi(r_N^*).$$

The repeated averaging shows that $E\zeta_{\varepsilon,1} = u_\varepsilon(r_0)$. Applying a similar expression for the variance, we also can show that $D\zeta_{\varepsilon,1}$ is bounded due to $g(r)$ being bounded.

If we set $c = 0$ in the input differential equation, we obtain the Poisson equation. In this case

$$G(r') = \frac{1}{4\pi}\left(\frac{1}{|r-r'|} - \frac{1}{d(r)}\right), \quad |r - r'| \leq d(r) = d,$$

$$f_\rho(x) = \frac{6x(1 - x/d)}{d^2}, \quad 0 \leq x \leq d.$$

Setting $y = x/d$, we obtain

$$f_\rho(y) = 6y(1-y), \quad 0 \leq y \leq 1. \tag{3.15}$$

Expression (3.15) corresponds to the distribution of the second order statistics of three sampled values of a random variable uniformly distributed in [0,1].

Now let us estimate the number R_ε of arithmetic operations required to obtain a prescribed error ε in estimating the solution. It has been shown previously that the number of spheres, when the trajectories are terminated in Γ_ε, is equal to $|\ln\varepsilon|$ by the order of magnitude. On the other hand, for the probabilistic error to be of the order of ε the number of the trajectories to be modelled should be $C\varepsilon^{-2}$. Hence,

$$R_\varepsilon \sim C_n |\ln\varepsilon|/\varepsilon^2,$$

where C_n is the number of arithmetic operations per one sphere in an n-dimensional space. For large values of n the quantity C_n depends linearly on n.

In solving some applied problems (e.g. the problems of electron-ion optics) it is necessary to calculate the gradient of the potential, i.e. the derivatives of the solution of the problem

$$\Delta u = -g, \quad u\big|_\Gamma = \psi. \tag{3.16}$$

Dyad'kin and Starikov [12] obtained an expression of the 'weight factor' for estimating the solution to the Laplace equation at two points by common sampled walk on spheres trajectories. Using this idea we can construct an algorithm of direct estimation of the derivatives of the solution of problem (3.16) at the point $(x_0, y_0, z_0) = P_0 = r_0$.

Using the reasoning similar to the one given above, for the point $P_{0,x} = (x_0 + x, y_0, z_0)$ we obtain

$$u(P_{0,x}) = \int_{S(P_0)} p(\omega, x)u(s)ds + \int_{|r-P_0|<d_0} g(r)G_x(r, d_0)dr, \qquad (3.17)$$
$$x < d(P_0) = d_0.$$

The Green function $G_x(r, d_0)$ for a ball and its normal derivative $p(\omega, x)$ are expressed by the formulae:

$$G_x(r, d_0) = \frac{1}{4\pi}\Big(\frac{1}{|r - P_0, x|} - \frac{d_0}{(x^2r^2 + d_0^4 - 2d_0xx_1)^{1/2}}\Big)$$

$$p(\omega, x) = \frac{1}{4\pi}\frac{d_0(d_0^2 - x^2)}{(d_0^2 + x^2 - 2d_0xa)^{3/2}},$$

where ω is a unit vector pointing from P_0 to P_1 (i.e., $\omega = s/|s|$); $a = a(x, \omega)$ is the cosine of the angle between ω and the x-axis; $x_1 = x(r - P_0)$. By taking the derivative of (3.17) with respect to x we obtain the relationship (see [13]):

$$\begin{aligned}\frac{\partial u}{\partial x}(x_0, y_0, z_0) &= \frac{1}{4\pi}\int_{|r-P_0|<d_0} \frac{x_1(d_0^3 - |r - P_0|^3)}{d_0^3 - |r - P_0|^3} g(r)dr \\ &+ \mathrm{E}\Big\{\frac{3a}{d_0}\Big[\sum_{i=1}^{N} \frac{1}{4\pi d_i}\int_{|r-P_i|<d} \frac{d_i - |r - r_i|}{|r - r_i|} g(r)dr + \psi(r_N^*)\Big]\Big\}.\end{aligned} \qquad (3.18)$$

The method considered also allows higher derivatives to be estimated.

It is easily seen that relationship (3.17) enables one to estimate the solution at all the points of the ball $|r - P_0| \leq d_0$ using the same trajectories in the chain $P_0, r_1, r_2, \ldots$. It is helpful for two reasons: first, thereby we save computer time and, second, statistical estimates of the values of $u(r)$ thus obtained are dependent, which allows a more precise estimation of the function on the whole.

The first integral appearing in (3.18) can be estimated by the Monte Carlo method by the single random 'node' with the density

$$[f_x(r) + f_y(r) + f_z(r)]^{1/2} = \frac{d_0^3 - |r - P_0|^3}{d_0^3|r - P_0|^2}, \quad |r - P_0| < d_0 .$$

The estimates of derivatives with respect to the parameter c can be obtained by the means of taking the corresponding derivative of the quantity ζ_ε:

$$\frac{\partial \zeta_\varepsilon}{\partial c} = \sum_{n=0}^{N}\Big[\frac{\partial Q_n}{\partial c}h(r_n) + Q_n\frac{\partial h(r_n)}{\partial c}\Big].$$

In this case we have [13]

$$\frac{\partial Q_n}{\partial c} = \frac{1}{2}Q_n\Big(\frac{n}{c} - \sum_{i=0}^{n-1}\frac{d_i \mathrm{ch}(d_i\sqrt{c})}{\sqrt{c}\,\mathrm{sh}(d_i\sqrt{c})}\Big)$$

$$\frac{\partial h(z)}{\partial c} = \frac{1}{4\pi}\int_{D(r)} \frac{\mathrm{sh}[(d-|r'-r|)\sqrt{c}]}{|r'-r|\mathrm{sh}(d\sqrt{c})} g(r') \frac{1}{2\sqrt{c}}$$
$$\times\left(\frac{(d-|r'-r|)\mathrm{ch}[(d-|r'-r|)\sqrt{c}]}{\mathrm{sh}[(d-|r'-r|)\sqrt{c}]} - \frac{d\,\mathrm{ch}(d\sqrt{c})}{\mathrm{sh}(d\sqrt{c})}\right) dr'.$$

As pointed out in [37] (see also [39]) a numerically equivalent expression for $\partial\zeta_\varepsilon/\partial c$ can be obtained from a vector estimate of the solution to a triangular system of differential equations obtained by taking derivatives of equation (3.14) with respect to the parameter c:

$$\begin{aligned} u &= Ku + h\,, \\ u' &= Ku' + K'u + h'. \end{aligned} \tag{3.19}$$

The operator K' is defined by the factor

$$s'(c,d) = \frac{\partial}{\partial c}\left[\frac{d\sqrt{c}}{\mathrm{sh}(d\sqrt{c})}\right] = \frac{d}{2\sqrt{c}}\cdot\frac{1}{\mathrm{sh}(d\sqrt{c})}[\mathrm{sh}(d\sqrt{c}) - d\sqrt{c}\,\mathrm{ch}(d\sqrt{c})]\;,$$

which is bounded for $c \geq 0$ (see [37]). The function h' is also bounded, therefore, the variance of the estimate of the derivative is uniformly bounded in c [37,39]. It is easy to verify that the variances in ε are uniformly bounded. Let us represent the vector estimate of the solution to system (3.18) in the form:

$$\begin{aligned} \zeta_{r_0}^{(1)} &= s(c,d_0)\zeta_{r_1}^{(1)} + h(c,r_0) \\ \zeta_{r_0}^{(2)} &= s(c,d_0)\zeta_{r_1}^{(2)} + s'(c,d_0)\zeta_{r_1}^{(1)} + h'(c,r_0). \end{aligned} \tag{3.20}$$

As mentioned above, the quantity $\mathrm{E}(\zeta^{(1)})^2$ and the functions $s'(c,d_0)$, $h(c,r_0)$ and $h'(c,r_0)$ are uniformly bounded [with allowance made for the boundedness of the functions (ψ,ψ',g,g')]. Therefore, from (3.20) it easy to obtain the uniform boundedness of the quantity $\mathrm{E}(\zeta^{(1)}\zeta^{(2)})$ and also the quantity $\mathrm{E}(\zeta^{(2)})^2$ in the same way is was done for $\mathrm{E}\zeta_\varepsilon^2$. Evidently, the conclusions drawn can be extended to the derivatives of any order. This is how the estimates of the solution of the Rikye problem are constructed (see Section 4.1). For $g \equiv 0$ these estimates are equivalent to the vector estimates given in [47]. Note that the Cauchy formula used in [47] converges to the corresponding derivative of the solution with increasing the number of nodes of quadrature, i.e. the integral can be calculated exactly.

Note that Ermakov *et al.* [18] used the integral representation of problem (3.1) in the form of (3.14) for constructing an algorithm of walk on spheres (and balls D_i) for the case of a variable parameter $c(r) \leq 0$. Such an algorithm is easy to construct, if the quantity

$$(c_m + c)u, \quad |c(r)| \leq c_m$$

is subtracted from both sides of equation (3.1) and the corresponding changes are made in the integral over the ball in (3.14), with an unknown function appearing in the integral over the ball. Evidently, the spectral radius of the integral operator thus obtained is less than unity, and the above justification of the algorithm (including the uniform boundedness of the variance) is carried over to this case. In the next section we shall use a simpler version of the integral representation to construct the algorithm for $c(r) > 0$. This version is not a special case of the version in [18], as the value $c_m = 0$ will be used.

3.4. NEW ALGORITHMS FOR VARIABLE $c(r)$

Let us rewrite (3.1) in the form:

$$\Delta u = -cu - g, \quad u\big|_\Gamma = \psi. \tag{3.21}$$

From (3.21) it is clear that, by analogy with Section 3.1, the following integral equation can be written for the function $u(r)$:

$$u = Ku + h. \tag{3.22}$$

In $D\backslash\Gamma_\varepsilon$ this equation has the form:

$$u(r) = \frac{1}{4\pi d^2(r)}\int_{S(r)} u(r(s))ds + \int_{D(r)} c(r')G_r(r')u(r')dr' + \int_{D(r)} G_r(r')g(r')dr' ,$$

$$r \in D\backslash\Gamma_\varepsilon$$

while in Γ_ε it transforms into the identity:

$$u(r) \equiv u(r), \quad r \in \Gamma_\varepsilon,$$

i.e. $k(r,r') \equiv 0$ and $h(r) \equiv u(r)$ for $r \in \Gamma_\varepsilon$. Here

$$G_r(r') = \frac{1}{4\pi}\Big(\frac{1}{|r-r'|} - \frac{1}{d(r)}\Big), \quad |r-r'| \le d(r)$$

is the Green function for a ball at the center of the ball $D(r)$. For $r \in D\backslash\Gamma_\varepsilon$ we write the integral operator K in the form:

$$\begin{aligned}&[1-c_0d^2(r)/6]\frac{[1-c_0d^2(r)/6]^{-1}}{4\pi d^2(r)}\int_{S(r)} u(r(s))ds\\ &+\frac{c_0d^2(r)}{6}\int_{D(r)}\frac{c(r')}{c_0}6d^{-2}(r)G_r(r')u(r')dr'\end{aligned} \tag{3.23}$$

with c_0 satisfying the inequality

$$c_0d^2_{\max}/6 < 1.$$

As a standard Monte Carlo estimate implements a probabilistic representation of the Neumann series, for the construction of the estimate it is necessary to randomize the result of applying the integral operator K. According to (3.23), this is accomplished in the following way: with the probability $1-c_0d^2(r_i)/6$, a new point r_{i+1} is chosen uniformly over the sphere $S(r_i)$, while the weight Q_i is multiplied by the quantity

$$q(r_i, c_0) = [1-c_0d^2(r_i)/6]^{-1};$$

with the opposite probability $c_0d^2(r_i)/6$, a point r_{i+1} is chosen within the ball with respect to the density $6d^{-2}(r_i)G_{r_i}(r')$ (see Section 3.3), while the weight is multiplied by $c(r_{i+1})/c_0$. In modelling the chain and constructing the estimate, the neighbourhood is allowed for precisely in the same manner as in Sections 3.1–3, i.e. when the chain finds itself inside Γ_ε it is terminated, and the solution estimate multiplied by the weight is added to the sum in the counter. The resulting random estimate has the form:

$$\xi = \sum_{n=0}^{N} Q_n h(r_n),$$

where

$$Q_n = \left[\prod_{i=1}^{m_n} \frac{c(r_{k_i})}{c_0}\right]\left[\prod_{i=1}^{n-m_n} q(r_i, c_0)\right]$$

and the function $h(r)$ is calculated exactly or in a randomized manner, as pointed out in Section 3.3 for the Poisson equation. Here $\{r_{t_i}\}$, $i = 1, \ldots, n-m_n$, are the points chosen on the spheres, while $\{r_{k_i}\}$, $i = 1, \ldots, m_n$, are the points chosen inside the balls.

In the new walk on spheres and balls the average numbers of steps onto spheres is asymptotically equivalent to the total average number of steps, as $\mathrm{E}\sum d_i^2 < +\infty$, and, hence, the number of steps into the balls is finite with a probability of 1 (which fact has been mentioned in [18]). On the other hand, the average number of steps onto spheres does not exceed the average numbers of steps in the standard walk on spheres, since intermediate steps into the balls only increase the 'drift' toward the boundary. Hence, for the new walk on spheres the average modelling costs S_ε also admit a logarithmic estimate:

$$S_\varepsilon \leq \text{const}\,|\ln \varepsilon|.$$

Below we shall justify an asymptotical unbiasedness of the quantity ξ_ε considered as an estimate of the solution of the initial problem.

Lemma 3.1. *Provided*

$$c_0 \geq |c(r)|, \quad c_0 < 0.6079c^* \tag{3.24}$$

the Neumann series for equation (3.22) converges.

Proof. Using the expansion of the function $\sin x$ into a Maclaurin series it can easily be shown that the inequality

$$\frac{c_0}{c^*} < \frac{6}{x_0^2}\left(1 - \frac{\sin x_0}{x_0}\right) \tag{3.25}$$

entails the inequality

$$\frac{1}{1 - c_0 d_i^2/6} < \frac{d_i\sqrt{c^*}}{\sin(d_i\sqrt{c^*})}. \tag{3.26}$$

Here $x_0 = d_{\max}\sqrt{c^*} \leq \pi$. For $x_0 = \pi$, from (3.25) we have $c_0 < 0.6079c^*$. With allowance made for the inequality $|c(r)|/c_0 < 1$, relationship (3.26) implies that the quantity ξ_ε can be termwise majorized by the standard estimate ζ (see Section 3.2) on walk on spheres $\{r_i\}$ (not necessarily maximal) with a certain $c < c^*$ for $\varepsilon = 0$. The Neumann series corresponding to the estimate ζ converges, as ζ can be obtained by the partial averaging (see Section 3.2) of the probabilistic representation of the solution. □

Note that for special types of domains, unbiasedness condition (3.24) can be extended with allowance made for the exact value of $d_{\max}$. For example, for a cube of edge l we have

$$c^* = 3\pi^2/l^2, \quad d_{\max} = l/2, \quad c_0/c^* \leq 0.6888.$$

Lemma 3.2. *If $u_0(r)$ is a Neumann series and K_0 is an integral operator of equation (3.22) with $c \equiv c_0 > 0$, $c_0 < 0.6079c^*$ and $\psi \equiv g \equiv C_0 > 0$, then*

$$u_0(r) \geq C_0, \quad K_0^n u_0(r) \underset{n\to\infty}{\longrightarrow} 0.$$

Proof. The first statement of the lemma is evident, while the second one results from the representation

$$u_0 = \sum_{i-0}^{n} K_0^i h + K_0^{n+1} u_0$$

which is connected with the fact that under the hypotheses of the lemma the Neumann series converges and satisfies equation (3.22). □

Theorem 3.1. *Under the hypotheses of Lemma 3.1 there exists a unique bounded solution (3.22) that can be represented by the corresponding Neumann series and coincides with the solution of problem (3.21).*

Proof. If $u(r)$ is a solution to equation (3.22) and $|u(r)| < C_1$, then

$$u = \sum_{i=0}^{n} K^i h + K^{n+1} u, \quad |K^{n+1} u| < C_1 C_0^{-1} K_0^{n+1} u_0.$$

The last-mentioned statement of the theorem is evident. □

Now let us proceed with constructing and studying the actual estimate ξ, which is obtained (see also Section 3.3) by the replacement of $u(r)$ in Γ_ε with the value of the boundary function ψ at the point $P(r) \in \Gamma$ closest to r. In this case

$$h(r) = \psi(P(r)), \quad r \in \Gamma_\varepsilon.$$

Theorem 3.2. *If the first derivatives of the solution of problem (3.1) are bounded in D, then, provided the conditions of Lemma 3.1 are satisfied, $\mathrm{E}\,\xi_\varepsilon = u_\varepsilon$ exists and*

$$|u(r) - u_\varepsilon(r)| \leq \text{const} \cdot \varepsilon, \quad r \in D, \quad \varepsilon > 0.$$

Proof. Evidently, the following relations are valid:

$$\xi - \xi_\varepsilon = Q_N[u(r_n) - \psi(P(r_N))], \quad |\xi - \xi_\varepsilon| \leq \text{const} \cdot Q_N^{(0)} \varepsilon\,,$$

where, similar to the proof of Lemma 3.1, $Q_N^{(0)}$ is the weight corresponding to a certain standard estimate (see Section 3.2) on walk on spheres $\{r_i\}$ with $c < c^*$ and $g \equiv 0$. □

Theorem 3.1a. *Under the conditions*

$$|c(r)| < c^*, \quad c_0 < 0.6079c^*$$

there exists a unique bounded solution to equation (3.22) that can be represented by the corresponding Neumann series and coincides with the solution of problem (3.21).

Proof. As was mentioned above, the number of steps into the balls is finite with probability 1. Because of that in the case $\psi \equiv c \equiv$ const the function $\mathrm{E}\,\xi$ is represented by the power series with the positive coefficients, which for sufficiently small values c, evidently, coincides with the similar power series, which represents the corresponding solution u. Hence, for the constant positive values ψ and c Theorem 3.1a holds. Because of that it is possible to prove Lemma 3.1 under the condition $|c(r)| < c^*$, on the basis of the following majorizing: $|c(r)| < c_1 < C^*$. □

Theorem 3.2a. *If the first derivatives of the solution of problem (3.1) are bounded in D, then, provided the conditions of Theorem 3.1a are satisfied, $\mathrm{E}\,\xi_\varepsilon = u_\varepsilon$ exists and*

$$|u(r) - u_\varepsilon(r)| \leq \text{const} \cdot \varepsilon, \quad r \in D, \quad \varepsilon > 0.$$

Proof. If $\psi \equiv 1$ and $c \geq 0$, then, provided the conditions of Theorem 3.1a are satisfied, we have

$$\mathrm{E}\,|Q_N| \leq \mathrm{E}\,|Q_N u(r_N)| < +\infty\,.$$

Therefore

$$|u(r) - u_\varepsilon(r)| = \mathrm{E}\,|Q_N[u(r_N) - \psi(P(r_N))]| \leq \text{const} \cdot \varepsilon.$$

□

Theorem 3.3. *If the conditions*

$$c_0 \geq |c(r)|, \quad c_0 \leq 0.488c^* \tag{3.27}$$

are satisfied and $g \equiv 0$, then $\mathrm{D}\xi_\varepsilon < C_d < +\infty$ for all $\varepsilon > 0$.

Proof. Direct calculations show that the quantity

$$\max\left\{y > 0 : \frac{t}{\sin t}\left(1 - \frac{y}{6}t^2\right)^2 \geq 1\right\}$$

is attained at a point y^* such that $0.488 < y^* < 0.489$, the quantity in braces being minimal and equal to 1 at $t \approx 2.175$. This means that the quantity ξ_ε^2 here can be majorized in the same way as ξ in the proof of Lemma 3.1. □

Now let us consider the variance for $g \not\equiv 0$. With the aim to extend the condition of the finiteness of the variance, let us modify the estimate as follows. We shall calculate the function $h(r)$ only for the step $r \to r'$ onto the sphere, however, with the weight $q(r, c_0)$, i.e. $h(r) \to h_1(r)$:

$$h_1(r_{k_i-1}) = 0, \quad h_1(r_{t_i-1}) = q(r_{t_i-1})h(r_{t_i-1}).$$

Thus, instead of the estimate ξ_ε we shall consider the estimate

$$\xi_{\varepsilon,1} = \sum_{n=0}^{N} Q_n h_1(r_n) = \sum_{i=1}^{N-m_N} Q_{t_i} h(r_{t_i-1}) + Q_N \psi(P(r_N)).$$

The repeated averaging yields: $\mathrm{E}\,\xi_{\varepsilon,1} = \mathrm{E}\,\xi$.

Theorem 3.4. *Under the hypotheses of Theorem 3.3 we have*

$$\mathrm{D}\,\xi_{\varepsilon,1} < C_{d,1} < +\infty \quad \forall \varepsilon > 0.$$

Proof. Similar to the proof of Lemma 3.1, the quantity $\xi_{\varepsilon,1}$ is termwise majorized by the standard estimate ζ for $c \equiv c_0 < 0.488c^*$ with the replacement $h \to |h_1|$. In doing so each factor of the form of c/c_0 is replaced by 1 and

$$Q_N^{(0)} - 1 = \sum_{n=1}^{N} (Q_n^{(0)} - Q_{n-1}^{(0)}) = \sum_{i=1}^{N-m_n} (Q_{t_i}^{(0)} - Q_{t_i-1}^{(0)}) = \frac{c_0}{6} \sum_{i=1}^{N-m_n} Q_{t_i} d_{t_i-1}^2.$$

Hence, here

$$|\zeta_{\varepsilon,1}| \le \zeta < \mathrm{const} \cdot Q_N^{(0)} .$$

□

Now let us construct estimates of the derivatives

$$u^{(k)}(r, c) = \frac{\partial^k u(r, c)}{\partial c^k}$$

while assuming $c = \mathrm{const}$, $|c| < c_0$. Evidently, it is sufficient to consider the case when the functions g and ψ are independent of c. Let us differentiate the estimate $\xi_{\varepsilon,1}$ k-times with respect to c to get:

$$\xi_{\varepsilon,1}^{(k)} = \frac{\partial^k \xi_{\varepsilon,1}}{\partial c^k} = \sum_{i=1}^{N-m_n} \frac{\mu_i!}{(\mu_i - k)!} c^{-k} Q_{t_i} h(r_{t_i-1}) + \frac{m_N!}{(m_N - k)!} c^{-k} Q_N \psi(P(r_N)).$$

Here μ_i is the number of steps into the balls among the first t_i steps, i.e. $\mu_i = m_{t_i}$. If $\mu_i < k$, then the corresponding term equals zero. Since

$$q_i = \frac{\mu_i!}{(\mu_i - k)!} \left(\frac{c}{c_0}\right)^{\mu_i} \to 0, \quad \mu_i \to \infty$$

the quantities q_i are uniformly bounded in i. Therefore, by analogy with Theorem 3.4, we can state that $\mathrm{D}\,\xi_{\varepsilon,1}^{(k)}$ is bounded uniformly in ε.

If the functions $u^{(k)}(r)$ are uniformly bounded, the estimate $\xi_{\varepsilon,1}^{(k)}$ is ε-biased, since

$$|u_\varepsilon^{(k)} - u^{(k)}| \le \mathrm{E}\,|Q_N^{(k)}[\psi(P(r_N)) - u(r_N, c)]| + \mathrm{E}\,|Q_N u^{(k)}(r_N, c)|.$$

Note that the estimates of derivatives simultaneously provides us with the estimates of the solution of the Dirichlet problem for a metaharmonic equation obtained from (3.1) by direct differentiation with respect to the scalar parameter (see Section 4.1).

Similar estimates can be also constructed for a piecewise constant parameter $c(r)$ whose individual values vary. These estimates can be used for solving the corresponding reverse and stochastic problems (see Sections 2.2 and 5.8).

Especially simple with the new algorithm is estimation of the derivatives of the solution by coordinates with making use of the expressions given in Section 3.3 for the corresponding derivatives of the non-central Green function for a ball and its normal derivative on the sphere. If the first step is onto a sphere, the additional algorithm here exactly coincides with that in Section 3.3, otherwise, the weight is multiplied by the ratio of the derivative of the Green function for a ball (see Section 3.3) to the transition density used for the first sphere. The coordinate derivatives of the function $u_\varepsilon^{(k)}$ are constructed similarly. Thus, such estimates possess the ε-biasedness property.

The new algorithms have been used for calculating a number of versions of problem (3.12) with $\varepsilon = 0.001$, $x = y = z = 0.5$, $N = 100000$. The results are listed in Table 3.4.

Table 3.4. Results of calculations by the new method for problem (3.12)

c	c_0	$(y_N \pm \sigma_N) \times 10^3$	σ^2	$u'_\varepsilon \times 10^3$	$u''_\varepsilon \times 10^3$	$u'''_\varepsilon \times 10^3$
14	14.1	437 ± 4.3	1.81	38.5 ± 8.7	5.04 ± 0.29	0.90 ± 0.13
1	1.1	178 ± 1.3	0.157	10.1 ± 0.32	0.87 ± 0.13	0.049 ± 0.049
1	10	174 ± 1.8	0.310	10.3 ± 0.14	0.81 ± 0.02	0.089 ± 0.004

The results presented are in good agreement with the results for $c = 14$ obtained by the standard walk on spheres method (see Section 3.2). The variance here is even somewhat smaller, which is likely to be accounted for by the presence of the weight factor of the form of $(c/c_0)^{m_n}$. On the other hand, the costs of the new method are practically the same as those of the standard one, since, though the average number of spheres increased by 13%, this increase is compensated for by the fact that it is not necessary to calculate the function $\sin x$ at each step.

It is interesting to note that for small values of c the derivatives are much better estimated with relatively large values of c_0. For the version with $c = 14$ and $c_0 = 14.1$, in addition to those listed in Table 3.4, the following derivatives have been also calculated:

$$u_\varepsilon^{(4)} \approx (0.214 \pm 0.063) \times 10^{-3}, \quad u_\varepsilon^{(5)} \approx (0.064 \pm 0.029) \times 10^{-3}.$$

On the basis of these estimates of the solution and on the basis of the first five derivatives, the values of the solution for $c = 16, 17, 19$ have been estimated using the Taylor series with the aim to compare with the corresponding results from Section 3.2. The mean-square error was estimated by the formula

$$\sigma\Big(\sum_{i=1}^{n} \xi_i\Big) \le \sum_{i=1}^{n} \sigma(\xi_i) = \tilde{\sigma}.$$

The results are included in Table 3.5.

For convenience of the comparison, the mean-square errors from Section 2 have been recalculated to a sampling volume of $N = 100000$. Table 3.5, with allowance made for the obvious overestimating of $\tilde{\sigma}$, demonstrates a high efficiency of the method of extrapolation in c.

Table 3.5. Estimates with the Taylor series at the point c=14

c	$y_N \pm \tilde{\sigma}_N$	Estimates from Section 3.2
14	0.437 ± 0.0043	0.438 ± 0.0044
16	0.525 ± 0.0068	0.528 ± 0.0068
17	0.580 ± 0.0090	0.583 ± 0.0095
19	0.718 ± 0.017	0.726 ± 0.018

Note that the random numbers α were simulated here by the method of residues by the formula:

$$u_0 = 1, \quad u_{n+1} \equiv u_n 5^{17} (\mathrm{mod}\, 2^{40}), \quad \alpha_n = u_n 2^{40}.$$

3.5. 'WALK ON SPHERES' ALGORITHMS FOR SOLVING HELMHOLTZ EQUATION IN THE n-DIMENSIONAL EUCLIDEAN SPACE

In this Section the algorithms of Monte Carlo methods for solving the n-dimensional Helmholtz equation are investigated. The dependence of the computational efficiency of the algorithms on n is studied [7].

Let us consider a n-dimensional Dirichlet problem for a variable-parameter Helmholtz equation

$$\Delta u + cu = -g, \quad u\big|_{\Gamma} = \psi, \tag{3.28}$$

as a main problem in a domain $D \in R^n$ with the boundary Γ, where $c(r) < c^*$, and $-c^*$ is the first eigenvalue of the Laplasian operator for the domain D, $r = (x_1, \dots, x_n) \in D$.

The conditions for the functions c, g, ψ and Γ to be regular are assumed to be fulfilled. These conditions guarantee the existence and uniqueness of the solution of problem (3.28), as well as its probabilistic representation and integral representation with making use of the Green function for a n-dimensional ball (see, e.g., [13, 18]).

The estimates to be considered below are associated with the n-dimensional 'walk on spheres' process within the domain D. To describe the process, let us introduce the following notation: $\bar{D}$ is a closure of domain D; $d(P)$ is a distance from the point P to the boundary Γ; Γ_ε is an ε-neighbourhood of the boundary Γ, i.e.,

$$\Gamma_\varepsilon = \{P \in \bar{D} : \ d(P) < \varepsilon\};$$

$S(P)$ is the largest n-dimensional sphere of those centred at the point P and entirely lying in $\bar{D}$, i.e.,

$$S(P) = \{Q \in \bar{D} : \ |Q - P| = d(P)\}.$$

In the 'walk on spheres' process, every next point P_{k+1} is chosen uniformly over the surface of the sphere $S(P_k)$; the process terminates when a point find itself inside Γ_ε.

Let us denote by $S_0(P, \varepsilon)$ the surface of that part of the sphere $S(P)$, that belongs to the set Γ_ε. Construct a sphere S_ε with radius ε centred at the point of contact of the boundary Γ with the sphere $S(P)$. Then the area of the part of the sphere $S(P)$, entirely lies inside S_ε equals

$$\frac{(\sqrt{\pi})^{n-1}}{\Gamma((n-1)/2)(n-1)} \varepsilon^{n-1}.$$

This yields the following lower bound for the probability of the next point ending up inside Γ_ε:

$$\frac{S_0(P,\varepsilon)}{S(P)} \geq \frac{\Gamma(n/2)}{\pi\Gamma((n-1)/2)(n-1)} \cdot \frac{\varepsilon^{n-1}}{d_{\max}^{n-1}} = \nu(\varepsilon), \tag{3.29}$$

where $d_{\max}$ is the exact upper boundary of radii of the spheres entirely lying in D. Evidently, the mean number $\mathrm{E}N = q(P,\varepsilon)$ of the transitions in the 'walk on spheres' chain, which governs the average time it takes for a computer to model the chain, does not exceed $\nu^{-1}(\varepsilon)$.

Moreover, for a wide class of boundaries Γ, the following logarithmic estimate is true [13,18]:

$$q(P,\varepsilon) \leq C_n |\ln \varepsilon|. \tag{3.30}$$

This estimate is obviously valid for convex domains. Muller [43] has shown that the mean number of the transitions in the 'walk on spheres' chain depends linearly on n for large values of n:

$$q(P,\varepsilon) \sim n. \tag{3.31}$$

For $c = \text{const} < c^*$, the probabilistic representation of the solution of problem (3.28) has the form [13]

$$u(r_0) = \mathrm{E}\int_0^\tau e^{ct} g(\xi(t))dt + E[e^{c\tau}\psi(\xi(\tau))],$$

where $\xi(t)$ is the n-dimensional diffusion process originating at the point r_0 and corresponding to the Laplasian operator, τ is the instant the process leaves the domain D for the first time. Proceeding from the process being strictly Markovian, we have

$$\begin{aligned} u(r_0) &= \mathrm{E}\sum_{i=0}^{\infty} \int_{\tau_i}^{\tau_{i+1}} e^{ct} g(\xi(t))dt + \mathrm{E}\left[\psi(\xi(\tau))\prod_{i=0}^{\infty} e^{c(\tau_{i+1}-\tau_i)}\right] \\ &= \sum_{i=0}^{\infty} \mathrm{E}\left[e^{c\tau_i}\int_0^{\tau_{i+1}-\tau_i} e^{ct} g(t+\tau_i)dt\right] + \mathrm{E}\left[\psi(\xi(\tau))\prod_{i=0}^{\infty} e^{c(\tau_{i+1}-\tau_i)}\right], \end{aligned}$$

where τ_i is the instant the process $\xi(t)$ for the first time arrives at the surface of the i-th sphere of the corresponding 'walk on spheres' $\{r_n\}$, $n = 0,\ldots,N$, and $r_N \in \Gamma_\varepsilon$.

On the other hand, the solution of problem (3.28) at the centre of the ball $D(r)$ can be represented as follows:

$$u(r) = \frac{s(c,d)}{\omega_n d^{n-1}} \int_{S(r)} u(r(s))ds + \int_{D(r)} G_r(r')g(r')dr',$$

where $G_r(r')$ is the Green function for a n-dimensional ball for the Helmholtz equation, ω_n is the area of the n-dimensional unit sphere

$$\omega_n = 2(\sqrt{\pi})^n/\Gamma(n/2),$$

$$s(c,d)=\begin{cases}\dfrac{(d\sqrt{c}/2)^{(n-2)/2}}{\Gamma(n/2)J_{(n-2)/2}(d\sqrt{c})}, & c>0,\\[2ex] \dfrac{(d\sqrt{c}/2)^{(n-2)/2}}{\Gamma(n/2)I_{(n-2)/2}(d\sqrt{c})}, & c\le 0.\end{cases}$$

Here $d\sqrt{c}<\alpha_n$, where α_n is the first positive root of the function $J_{(n-2)/2}(x)$. The latter requirement is evidently met, if $c<c^*$. Indeed, the quantity c^* decreases with expanding the domain, and for a ball D^* of radius $d_{\max}$ the relationship $d_{\max}\sqrt{c^*(D^*)}=\alpha_n$ is valid.

Let us denote by P the point of the boundary Γ closest to r_n. Using the repeated averaging procedure, we can easily obtain (see Section 3.2) that

$$u(r_0)\approx u_\varepsilon(r_0)=\mathrm{E}\eta_\varepsilon^{(0)},$$

where

$$\eta_\varepsilon^{(0)}=\sum_{i=0}^{N}\left[\prod_{j=0}^{i-1}s(c,d_j)\right]\int_{D(r_i)}G(\rho,c,d_i)g(\rho)d\rho+\left[\prod_{j=0}^{N-1}s(c,d_j)\right]\psi(P). \tag{3.32}$$

Here $d_j=d(r_j)$, $D(r_i)$ is the ball of radius d_i centred at the point r_i. If the first derivatives of the solution are bounded in $\bar{D}$, then

$$|u(r)-u_\varepsilon(r)|\le C\varepsilon,\quad r\in D.$$

Futher, each of the integrals appearing in (3.32) can be estimated by its own random node ρ_i. It is reasonable to choose random nodes in $D(r_i)$ with the density proportional to $G(\rho,c,d_i)$. In this case, instead of $\eta_\varepsilon^{(0)}$ we obtain the estimate

$$\eta_\varepsilon=\sum_{i=0}^{N}\left[\prod_{j=0}^{i-1}s(c,d_j)\right]g(\rho_i)F_{r_i}+\left[\prod_{j=0}^{N-1}s(c,d_j)\right]\psi(P),$$

where

$$F_r=\int_{D(r)}G(\rho,c,d(r))d\rho\le C_1d^2(r). \tag{3.33}$$

The repeated averaging shows that

$$\mathrm{E}\eta_\varepsilon=\mathrm{E}\eta_\varepsilon^{(0)}=u_\varepsilon(r_0).$$

Now let us consider the question of uniform in ε boundedness of the variance $\mathrm{D}\eta_\varepsilon$.

First, assume that $g\equiv 0$. In this case

$$\eta_\varepsilon\le\eta=C_2\left[\prod_{j=0}^{\infty}s(c,d_j)\right],$$

and $E\eta<+\infty$ for $c<c^*$. Hence,

$$\mathrm{E}\eta_\varepsilon^2\le C_2^2\mathrm{E}\left[\prod_{j=0}^{\infty}s^2(c,d_j)\right].$$

Now let us show that for $c<c^*$ the following inequality holds:

$$s^2(c/2, d) \leq s(c, d). \tag{3.34}$$

Evidently, that $s(c,d) = v(r)$, where $v(r)$ is the solution of the problem

$$\Delta v + cv = 0, \quad v|_{S(r)} = 1,$$

where $S(r)$ is the sphere with radius d centred at the point r. On the other hand, $v(r) = \mathrm{E}e^{c\tau}$, where τ is the instant the diffusion process originating at the point r leaves the ball $D(r)$. Hence,

$$s^2(c/2, d) = (\mathrm{E}e^{c\tau/2})^2 \leq \mathrm{E}e^{c\tau} = s(c, d),$$

and the uniform in ε bondedness of the variance $\mathrm{D}\eta_\varepsilon$ has been proved for $c < c^*/2$ at $g \equiv 0$.

Now let us turn to problem (3.28) for an inhomogeneous Helmholtz equation. Evidently, it will suffice to prove uniform boundedness of the variance for $g \equiv \text{const}$, and the statement is true, if for $\varepsilon = 0$ the variance of the quantity

$$\eta_n = \sum_{i=0}^{n} \left[\prod_{j=0}^{i-1} s(c, d_j) \right] d_i^2$$

is bounded uniformly in n.

Let us consider the relationship

$$\prod_{j=0}^{n} s(c, d_j) = 1 + \sum_{i=0}^{n} \left[\prod_{j=0}^{i-1} s(c, d_j) \right] [-1 + s(c, d_i)] = \zeta_n.$$

As $d_{\max}\sqrt{c} < \alpha_n$, then

$$s(c, d_i) - 1 \geq C_3 d_i^2, \quad \eta_n < C_3^{-1} \zeta_n.$$

From the above we have

$$\lim_{n \to \infty} \mathrm{D}\zeta_n < +\infty,$$

and the quantity $\mathrm{D}\eta_n$ possesses the same property.

Now let us estimate the mean number $R_{\varepsilon,n}$ of arithmetic operations, required to obtain a prescribed error ε in estimating the solution. Let us note that

$$R_{\varepsilon,n} \sim L_\varepsilon Q_{\varepsilon,n} C_n,$$

where L_ε is the number of the trajectories to be modelled (it should be equal $C\varepsilon^{-2}$ for the probabilistic error to be of the order of ε), $Q_{\varepsilon,n}$ is the mean number of spheres (it is equal to $n|ln\varepsilon|$ by the order of magnitude), C_n is the number of arithmetic operations per one sphere in an n-dimensional space (for large values of n it depends linearly on n [17]). Hence,

$$R_{\varepsilon,n} \sim n^2 |\ln \varepsilon| / \varepsilon^2.$$

Let us dwell on the results of methodical calculations. The Monte Carlo method was used to solve the Dirichlet problem in a cube $0 \leq x_i \leq 1, \quad i = 1, \ldots, n$ for the equation

$$\Delta u + cu = 0, \quad u|_\Gamma = \cos(x_1\sqrt{c/n}) \times \ldots \times \cos(x_n\sqrt{c/n}). \tag{3.35}$$

The exact solution has the form

$$u(r) = \cos(x_1\sqrt{c/n}) \times \ldots \times \cos(x_n\sqrt{c/n}).$$

It is known that for a n-dimensional unit cube $c^* = n\pi^2$. The solution was estimated for different values of c, for $n = 4$ at the point with the coordinates $x_1 = \ldots = x_n = 0.9$, with $\varepsilon = 0.0001$, $N = 40000$, $c^* \approx 39.5$ (Table 3.6), and for $n = 10$ at the point $x_1 = \ldots = x_n = 0.1$, with $\varepsilon = 0.01$, $N = 10000$, $c^* \approx 98.7$ (Table 3.7). Here y_n is the estimate of solution, σ_n is the estimate of the corresponding mean-square error and σ^2 is the estimate of the quantity $D\eta_\varepsilon$.

Table 3.6. Results of calculation for problem (3.35) with $n = 4$, $N = 40000$

c	$u(r,c)$	$y_n \pm \sigma_n$	σ^2
15	0.00086	0.00096 ± 0.00020	0.00168
20	0.03338	0.03347 ± 0.00025	0.00253
25	0.15571	0.15490 ± 0.00120	0.05722
30	0.36931	0.36357 ± 0.00631	1.59144
35	0.61982	0.59343 ± 0.02730	29.8515
40	0.83753	0.73218 ± 0.10720	459.666
45	0.97017	0.97484 ± 0.39973	6391.475

Table 3.7. Results of calculation with $n = 10$, $N = 10000$

c	$u(r,c)$	$y_n \pm \sigma_n$	σ^2
30	0.860	0.859 ± 0.0005	0.003
50	0.777	0.776 ± 0.0007	0.006
100	0.601	0.600 ± 0.0010	0.014
120	0.542	0.541 ± 0.0013	0.016

The analysis of results shows that insignificant deviations from the exact solution have been obtained for $c > c^*/2$ as well.

The following problem was also solved in the unit cube:

$$\Delta u + cu = 0, \quad u|_\Gamma = 1. \tag{3.36}$$

The solution was estimated at the point $x_1 = \ldots = x_n = 0.5$, for $n = 4$, $\varepsilon = 0.001$, $N = 40000$ (Table 3.8), and for $n = 10$, $\varepsilon = 0.01$, $N = 10000$ (Table 3.9). There exists no exact solution of this problem, but the results agree satisfactorily with the solution of problem (3.36) for the ball of radius d, where d is chosen in such a way that the quantity c^* for the ball coincides with the quantity c^* for the cube, i.e., $d = \alpha_n/\pi\sqrt{n}$. In Tables 3.8, 3.9 $s(c,d)$ is the exact solution of problem (3.36) at the centre of the ball.

It is interesting to note that for the same values of y_n the quantity σ_n decreases with growing n. The results are listed in Table 3.10. The time it takes for the computer to model problem (3.36) for $n_1 = 4$ with $\varepsilon = 0.001$, $N = 10000$ is $t_1 =$ 5min.20sec., and for $n_2 = 10$ is $t_2 =$ 34min.21sec., i.e., $t_1/t_2 \approx (n_1/n_2)^2 = 0.16$. The result agrees with the estimate $R_{\varepsilon,n} \sim n^2$.

Algorithms for variable $c(r)$ are considered below. Let us rewrite (3.28) in the form

$$\Delta u = -cu - g, \quad u|_\Gamma = \psi. \tag{3.37}$$

Using the integral representation of the solution of the Poisson equation at the ball center,

Table 3.8. Results of calculation for problem (3.36) with $n = 4$, $N = 40000$

c	$y_n \pm \sigma_n$	σ^2	$s(c,d)$
18	2.7665 ± 0.0116	5.3890	2.7276
19	2.9703 ± 0.0143	8.1388	2.9268
20	3.1966 ± 0.0176	12.3927	3.1484
22	3.7313 ± 0.0217	29.5330	3.6743

Table 3.9. Results of calculation with $n = 10$, $N = 10000$

c	$y_n \pm \sigma_n$	σ^2	$s(c,d)$
33	2.9101 ± 0.0173	2.9919	2.8824
34	2.9937 ± 0.0194	3.7476	2.9876
50	5.6103 ± 0.1018	103.5363	5.5542
55	7.0724 ± 0.1318	173.8112	6.8959

Table 3.10. Results of calculation with $N = 10000$, $\varepsilon = 0.01$

n	c	$y_n \pm \sigma_n$	n	c	$y_n \pm \sigma_n$
2	11.0	2.9191 ± 0.0331	6	25.0	2.9281 ± 0.0270
3	15.6	2.9559 ± 0.0287	8	29.9	2.9934 ± 0.0207
4	19.3	2.9381 ± 0.0260	10	34.0	2.9937 ± 0.0194

we can write the following integral equation for the function $u(r)$:

$$u = Ku + h. \tag{3.38}$$

In $D \setminus \Gamma_\varepsilon$ this equation has the form

$$u(r) = \frac{1}{\omega_n d^{n-1}(r)} \int_{S(r)} u(r(s))ds + \int_{D(r)} c(r')G_r(r')u(r')dr' + \int_{D(r)} G_r(r')g(r')dr', \qquad r \in D \setminus \Gamma_\varepsilon ,$$

while in Γ_ε it transforms into the identity

$$u(r) \equiv u(r), \quad r \in \Gamma_\varepsilon ,$$

i.e., $k(r,r') \equiv 0$ and $h(r) \equiv u(r)$ for $r \in \Gamma_\varepsilon$. Here $k(r,r')$ is the kernel of the integral operator K.

$$G_r(r') = \frac{1}{(n-2)\omega_n}\left(\frac{1}{|r-r'|^{n-2}} - \frac{1}{d(r)^{n-2}}\right), \quad |r-r'| \le d(r),$$

is the Green function for a ball at the centre of the n-dimensional ball $D(r)$. For $r \in D\setminus\Gamma_\varepsilon$ we write the integral operator K in the form

$$[1 - c_0 d^2(r)/2n] \int_{S(r)} \frac{[1 - c_0 d^2(r)/2n]^{-1}}{\omega_n d^{n-1}(r)} u(r(s))ds + \frac{c_0 d^2(r)}{2n} \int_{D(r)} \frac{c(r')}{c_0} 2nd^{-2}(r)G_r(r')u(r')dr', \tag{3.39}$$

with c_0 satisfying the inequality $c_0 d^2_{\max}/2n < 1$.

The standard Monte Carlo estimate that implements a probabilistic representation of the Neumann series has the form (see Section 1.2):

$$\xi = \sum_{n=0}^{N} Q_n h(r_n), \quad u(r_0) = \mathrm{E}\xi.$$

According to (3.38), the following 'walk on spheres and balls' are accomplished: with a probability of $1 - c_0 d^2(r_i)/2n$, a new point r_{i+1} is chosen uniformly over the sphere $S(r_i)$, while the weight Q_i is multiplied by the quantity

$$q(r_i, c_0) = [1 - c_0 d^2(r_i)/2n]^{-1}$$

with an opposite probability $c_0 d^2(r_i)/2n$, a point r_{i+1} is chosen within the ball with respect to the density $2nd^{-2}(r_i)G_{r_i}(r')$, while the weight is multiplied by $c(r_{i+1})/c_0$. When the chain find itself inside Γ_ε it is terminated, and the solution estimate multiplied by the weight is added to the sum in the counter. The resulting random weights Q_n has the form

$$Q_n = \left[\prod_{i=1}^{m_n} \frac{c(r_{k_i})}{c_0}\right] \left[\prod_{i=1}^{n-m_n} q(r_{t_i-1}, c_0)\right],$$

and the function $h(r)$ is calculated exactly or in a randomized manner. Here $\{r_{t_i}\}$, $i = 1, \ldots, n - m_n$ are the points chosen on the spheres, while $\{r_{k_i}\}$, $i = 1, \ldots, m_n$ are the points chosen inside the balls.

For the new 'walk on spheres and balls' the average modelling costs $S_{\varepsilon,n}$ also admit an estimate (see Section 3.4)

$$S_{\varepsilon,n} \leq \mathrm{const} |\ln \varepsilon| n^2.$$

Below we shall justify an asymptotical unbiasedness of the quantity ξ_ε considered as an estimate of the solution of the initial problem.

Lemma 3.3. *Provided*

$$c_0 \geq |c(r)| \quad \text{and} \quad c_0 < 2nc^*/\alpha_n^2, \tag{3.40}$$

the Neumann series for equation (3.38) converges.

Proof. Let us consider c_0 satisfying the inequality

$$\frac{c_0}{c^*} < \min_{x \in [0, \alpha_n]} \left\{ \frac{2n}{x^2} \left(1 - \frac{\Gamma(n/2) J_{(n-2)/2}(x)}{(x/2)^{(n-2)/2}} \right) \right\}, \tag{3.41}$$

where $x = d\sqrt{c^*}$, than for all $d_i \leq d_{\max}$

$$\frac{c_0}{c^*} < \frac{2n}{d_i^2 c^*} \left(1 - \frac{\Gamma(n/2) J_{(n-2)/2}(d_i \sqrt{c^*})}{(d_i \sqrt{c^*}/2)^{(n-2)/2}} \right),$$

that entails the inequality

$$\frac{1}{1 - c_0 d_i^2/2n} < \frac{(d_i\sqrt{c^*}/2)^{(n-2)/2}}{\Gamma(n/2)J_{(n-2)/2}(d_i\sqrt{c^*})} \,. \tag{3.42}$$

The minimized function monotone decreases in $[0, \alpha_n]$, hence, inequality (3.41) is equivalent to $c_0 < 2nc^*/\alpha_n^2$. With allowance made for inequality $|c(r)|/c_0 < 1$, relationship (3.42) implies that quantity ξ_ε can be termwise majorized by the standard estimate η on 'walk on spheres' $\{r_i\}$ (not necessarily maximal) with a certain $c < c^*$ for $\varepsilon = 0$. The Neumann series corresponding to estimate η converges, as η can be obtained by the partial averaging of the probabilistic representation of the solution. □

Let us note that for large n quantity α_n depends linearly on n, hence, $2n/\alpha_n^2 \to 0$ for $n \to \infty$. For $n = 3$ the Neumann series converges with $c_0 < 0.6079c^*$, for $n = 4$ with $c_0 < 0.5454c^*$, for $n = 10$ with $c_0 < 0.3481c^*$.

In Section 3.4 it is proved that under the hypotheses of Lemma 3.3 there exists a unique bounded solution to equation (3.38) that can be represented by the corresponding Neumann series and coincides with the solution of problem (3.37).

Now let us proceed with constructing and studying the actual estimate ξ, which is obtained by the replacement of $u(r)$ in Γ_ε with the value of the boundary function ψ at the point $P(r) \in \Gamma$ closest to r. In this case

$$h(r) = \psi(P(r)), \quad r \in \Gamma_\varepsilon \,.$$

If the first derivatives of the solution of problem (3.28) are bounded in D, then, provided the conditions of Lemma 3.3 are satisfied, $\mathrm{E}\xi_\varepsilon = u_\varepsilon$ exists (see Section 3.4) and

$$|u(r) - u_\varepsilon(r)| \leq \text{const} \cdot \varepsilon, \quad r \in D, \quad \varepsilon > 0.$$

Lemma 3.4. *If the conditions*

$$c_0 \geq |c(r)|, \quad \frac{c_0}{c^*} < \min_{x\in[0,\alpha_n]} \left\{ \frac{2n}{x^2}\left(1 - \left(\frac{\Gamma(n/2)J_{(n-2)/2}(x)}{(x/2)^{(n-2)/2}}\right)^{1/2}\right)\right\} \tag{3.43}$$

are satisfied and $g \equiv 0$, then $\mathrm{D}\xi_\varepsilon < c_d < +\infty$ for all $\varepsilon > 0$.

Proof. Relationship (3.43) implies that

$$(1 - c_0 d_i^2/2n)^{-2} < \frac{(d_i\sqrt{c^*}/2)^{(n-2)/2}}{\Gamma(n/2)J_{(n-2)/2}(d_i\sqrt{c^*})} \,.$$

This means that the quantity ξ_ε^2 here can be majorized in the same way as ξ in the proof of Lemma 3.4. □

Direct calculations show that the variance is bounded for $n = 3$ with $c_0 < 0.488c^*$, for $n = 4$ with $c_0 < 0.465c^*$, for $n = 10$ with $c_0 < 0.336c^*$.

Now let us consider the variance for an inhomogeneous equation. With the aim to extend the condition of the finiteness of the variance, instead of the estimate ξ_ε we shall consider the estimate

$$\xi_{\varepsilon,1} = \sum_{n=0}^{N-m_N} Q_n h_1(r_n) = \sum_{i=1}^{N-m_n} Q_{t_i} h(r_{t_i-1}) + Q_N \psi(P(r_N)).$$

The repeated averaging yields $E\xi_{\varepsilon,1} = E\xi$.

Under the hypotheses of Lemma 3.4 we have

$$D\xi_{\varepsilon,1} < C_{d,1} < +\infty \quad \forall \varepsilon > 0.$$

Now let us construct estimates of the derivatives

$$u^{(k)}(r,c) = \frac{\partial^k u(r,c)}{\partial c^k} ,$$

while assuming $c = \text{const} < c_0$. Evidently, it is sufficient to consider the case when the functions g and ψ are independed of c. Let us differentiate the estimate $\xi_{\varepsilon,1}$ k-times with respect to c to get

$$\xi_{\varepsilon,1}^{(k)} = \frac{\partial^k \xi_{\varepsilon,1}}{\partial c^k} = \sum_{i=1}^{N-m_N} \frac{\mu_i!}{(\mu_i - k)!} c^{-k} Q_{t_i} h(r_{t_i-1}) + \frac{m_N!}{(m_N - k)!} c^{-k} Q_N \psi(P(r_N)).$$

Here μ_i is the number of steps into the balls among the first t_i steps, i.e., $\mu_i = m_{t_i}$. If $\mu_i < k$, then the corresponding term equals zero. Therefore, by analogy with Lemmas 3.3,4 we can state that $E\xi_{\varepsilon,1}^{(k)}$ exists and the quantity $D\xi_{\varepsilon,1}^{(k)}$ is bounded uniformly in ε.

The new algorithms have been used for calculating a number of versions of problem (3.36) at the point $x_1 = \ldots = x_n = 0.5$ for $n = 4$ with $\varepsilon = 0.001$, $N = 40000$ and for $n = 10$ with $\varepsilon = 0.01$, $N = 10000$ (Table 3.11).

Table 3.11. Results of calculations by the new method for problem (3.36)

n	c	c_0	$(y_n \pm \sigma_n) \cdot 10^3$	σ^2	$u'_\varepsilon \cdot 10^3$	$u''_\varepsilon \cdot 10^3$	$u'''_\varepsilon \cdot 10^3$
4	18	18.1	2761 ± 11.8	5.568	189.5 ± 2.4	18.19 ± 0.56	2.22 ± 0.14
10	33	33.1	2939 ± 18.1	3.299	107.2 ± 2.1	4.83 ± 0.25	0.26 ± 0.03

The results presented are in good agreement with the results obtained by the standard 'walk on spheres' method. The costs of the new method are practically the same as those of the standard one.

On the basis of these estimates of the solution and on the basis of the first five derivatives using the Taylor series in the point $c = 18$ for $n = 4$ and $c = 33$ for $n = 10$, the values of the solution have been estimated for other values of c with the aim to compare with the corresponding results. The mean-square error was estimated by the formula

$$\sigma\left(\sum_{i=1}^{n} \xi_i\right) \le \sum_{i=1}^{n} \sigma(\xi_i) = \tilde{\sigma}.$$

The results are included in Tables 3.12, 13.

Table 3.12. Estimates with the Taylor series at the point $c = 14$ for $n = 4$

c	$y_n \pm \sigma_n$	Estimates from the table 3.8
18	2.761 ± 0.0118	2.767 ± 0.0116
19	2.960 ± 0.0145	2.970 ± 0.0143
20	3.180 ± 0.0180	3.197 ± 0.0176
22	3.692 ± 0.0280	3.731 ± 0.0217

Table 3.13. Estimates with the Taylor series at the point $c = 33$ for $n = 10$

c	$y_n \pm \sigma_n$	Estimates from the table 3.9
33	2.939 ± 0.0182	2.910 ± 0.0173
34	3.049 ± 0.0204	2.994 ± 0.0194
50	5.737 ± 0.1271	5.610 ± 0.1018
55	7.117 ± 0.2153	7.072 ± 0.1318

3.6. SOLVING DIFFERENCE EQUATIONS BY THE MONTE CARLO METHOD

In principle, it is possible to solve by the Monte Carlo method an arbitrary difference equations system after reducing it to the special form:

$$u = Au + f, \quad \rho(A) < 1,$$

where $\rho(A)$ is the spectral radius of the matrix A (see [9,17]). It is especially easy to represent by this form the conventional grid approximations for the standard elliptic equations. The corresponding matrices A are stochastic (or substohastic) and related to the symmetric 'walk-on-grid' [1].

Let us consider the first boundary value problem for the Poisson equation:

$$\frac{\partial^2 u}{\partial x^2} + \frac{\partial^2 u}{\partial y^2} = -g(x,y) \tag{3.44}$$

in the unity square with the boundary condition $u\big|_\Gamma = \psi(x,y)$, where Γ is the square boundary.

Using the second divided differences to approximate the partial derivatives in Eq. (3.44), we obtain the system of $(L-1)^2$ linear algebraic equations:

$$u_{i,j} = \frac{1}{4}(u_{i-1,j} + u_{i,j-1} + u_{i+1,j} + u_{i,j+1} + l^2 g_{i,j}), \quad 1 \le i,j \le L-1. \tag{3.45}$$

Here $u_{i,j}$ is the approximate value of $u(il, jl)$ inside the square. If $(il, jl)$is the boundary node, then $u(il, jl) = \psi(il, jl)$.

Besides, $g_{i,j} = g(il, jl)$. Relation of Eq. (3.45) has the form of the total average (if $g_{i,j} = 0$). Because of that, $u_{i,j}$ coincides with the average of the random variable $\xi_{i,j}$, which is constructed as follows:

a) the point is placed at the node (il, jl); the initial value of the counter equals $l^2 \cdot g_{i,j}/4$;

b) with the equal probabilities the point is placed at one of the neighbouring nodes and the corresponding value of $l^2 g/4$ is added to the counter.

Item (b) is repeated till the point comes to the boundary, where the trajectory is terminated and the corresponding value of ψ is added to the counter. The resulting value of the counter gives the sampling value of $\xi_{i,j}$. Obviously, the variables $\mathrm{E}\xi_{i,j}$ satisfy system of Eq. (3.45), i.e. $\mathrm{E}\xi_{i,j} = u_{i,j}$. Here the direct simulation is realized and the values $\mathrm{D}\xi_{i,j}$ are uniformly bounded according to Section 1.2.

For other elliptic equations, the algorithms are constructed similarly. For example, instead of Eq. (3.44) let us consider the equation

$$\Delta u - cu = -g.$$

Then in Eq. (3.45) the multiplier $1/4$ is changed by $1/(4+cl^2)$. To estimate $u_{i,j}$ in this case, it is possible to use the same symmetric walk-on-grid, introducing the auxiliary 'weight', which is multiplied by $(1+cl^2/4)^{-1}$ after each transition. The value $l^2g/4$ is changed by $l^2g/(4+cl^2)$ and added to the counter after multiplying by the weight.

If other boundary condition is given, then for constructing the algorithm it is necessary to consider the corresponding linear relations in the boundary nodes. For example, let the normal derivative $\partial u/\partial n$ equal to zero and one side of the square. Considering the difference approximation of this derivative we obtain the following change in the walk process: after coming to the mentioned boundary part, the point 'jumps' inside the region at the distance l.

Now let us estimate the average number R_ε of operations required to obtain the result with the error ε for m-dimensional problem, using the walk-on-grid. Firstly remark, that for the rectangular region the following relation holds:

$$|u_{i,j} - u(il, jl)| = O(l^2).$$

The similar relations hold for other regions, if boundary conditions are determined by linear interpolation. Therefore, to have the deterministic error of the order ε it is necessary to use the grid step size of the order $\sqrt{\varepsilon}$, i.e. $l = O(\sqrt{\varepsilon})$.

Now let us estimate the average number $M_k(l,m)$ of the transitions, preceding coming the point to the boundary, where m is the number of coordinates.

Clearly, the variables M_k satisfy system of Eq. (3.45) after changing the multiplier $1/4$ by $1/(2m)$, and the value l^2g - by $2m(\psi = 0)$.

Hence,

$$\Delta_l^{(m)} M = -\frac{2m}{l^2},$$

where $\Delta_l^{(m)}$ is the m-dimensional difference Laplace operator and M is the vector of the values M_k. It is known, that the operator $[\Delta_l^{(m)}]^{-1}$ is uniformly bounded with respect to l.

Therefore,

$$M_k(l,m) = \frac{c(m)}{l^2} + 0\left(\frac{1}{l^2}\right).$$

For the m-dimensional infinite plane slab $c(m) = c \cdot m$. So, if $l = O(\varepsilon)$, then

$$|u_{i,j} - u(il, jl)| = O(\varepsilon), \quad M \sim c(m) \cdot \varepsilon^{-1}.$$

On the other hand, to have the probabilistic error of the order ε, it is necessary to sample $c_1\varepsilon^{-2}$ trajectories.

Therefore, here

$$R_\varepsilon \sim \frac{c_1 c(m)}{\varepsilon^3} .$$

Let us consider the optimization problem for the 'walk-on-grid' method which is constructed using the standard finite difference approximation of the k-dimensional Dirichlet problem on the rectangular grid with the step size h:

$$\Delta_h u_h = -g_h, \quad u\Big|_{\Gamma_h} = \psi_h .$$

It is known that ψ_h can be constructed in the way ensuring the convergence of the linearly prolonged difference solution to the exact one with order $O(h^2)$. For the estimate u_h the isotropic walk-on-grid is carried out before reaching Γ_h. When passing each point the estimate is added to with the corresponding quantity $h^2 g_h/(2k)$, and when reaching Γ_h this estimate is added to with the quantity ψ_h. The mean value of such random estimate is equal to the value of u_h at the point which is the starting one for the walk (see, for example, [1]).

It is clear that the intermediate points of the trajectory of the walk-on-grid can be considered as initial ones for additional trajectories coinciding with the simulated one in other respects. The wide application and investigation of such 'global' walk-on-grid method have been described in [1]. Here, we consider some possible ways of theoretical optimization of such method.

Some authors have considered the walk starting at one of the intermediate points of the domain. In this case, it is obvious that the probability of transition to a given boundary point is of order h^{-k+1}. On the other hand, if the walk starts from the boundary, the average number of arrivals at different points, satisfying the difference Laplace equation, is one and the same if the uniform discrete distribution with probabilities of order h^{-k+1} is given at the boundary. Hence, in this case as well, the average number of first arrivals is estimated from below by the quantity of order h^{-k+1}, i.e. this order can be referred to different distributions of initial points.

The global method for the Laplace equation is realized in the simplest way: the numbers of the nodes passed are stored along the trajectory, and the summation is carried out when reaching the boundary. The cost of the trajectory in this case is inessentially superior to the cost of estimation of the solution at one point and on the average is defined by the quantity of the form $c(k)h^{-2}$, where $c(k)$ is a linearly increasing function asymptotically for $k \to \infty$ [17]. The optimization problem for the global method can thus be formulated here in the form

$$S = t_k N h^{-2} \to \min_{N,h}, \quad ch^4 + \frac{d}{Nh^{k-1}} = \delta^2 ,$$

where d is the upper bound of the estimate variance which is bounded uniformly in h. This problem can be formally reduced to the local estimate optimization problem (see Section 1.5) by making the replacement $N' = Nh^{k-1}, \quad k' = k+1$. The minimal cost is of order $\delta^{-(k+5)/2}$ here while in the simple point-by-point estimation it is estimated by the quantity of order $\delta^{-(k+6)/2}$. The results obtained can be easily generalized to the case of the Poisson equation by the use of the appropriate summarize method.

The Dirichlet problem can also be solved by the 'walk-on-grid' method using the conjugate representation which can be realized in a particularly simple way for the Laplace equation. In this case, the trajectories can be simulated with equal probabilities from different points of the boundary with the weight proportional to $\psi_h h^{-k+1}$ and the solution at this point can be estimated by the average weighted number of arrivals at this point. The simple analysis shows that the optimal cost is of the same order here as the corresponding quantity for the point-by-point estimation but it can be essentially superior to this quantity because of the sufficiently large value of variance (it can be easily seen for the case $\psi = \text{const}$).

Let us consider now some remarks on possible use of computers. It is first obvious that the use of M independent computational processors by distributing random samples among them decreases the cost of the Monte Carlo method by M times as the final summation and averaging of the results are practically inessential.

The conclusion of the previous section shows that the use of the simple parallel summator can essentially raise the efficiency of the 'walk-on-grid' method in solving difference equations. At the same time such device supplemented with a purely arithmetic parallel processor computing the values of the kernel can also essentially reduce the cost of the general method of local estimates in solving multidimensional integral equations (see Section 1.5).

3.7. ADDITIONAL REMARKS

1. The new algorithms for the variable non-negative parameter $c(r)$ extend the capabilities of the Monte Carlo method in solving metaharmonic equations, as well as parabolic and hyperbolic equations on the basis of Laplace and Fourier transforms [18, 47].

As indicated in [39], extra possibilities are offered by the randomization, which consists in solving a high-order equation successively with taking advantage of elementary random estimates of the solutions of primary problems. Of great importance is, of course, the problem linearity allowing the repeated averaging:

$$\mathrm{E}f(\xi,\eta) = \mathrm{E}\mathrm{E}[f(\xi,\eta)|\xi] \ . \tag{3.46}$$

To study the variance of the randomized estimate, the expression similar to the last-mentioned one is used:

$$\mathrm{D}f(\xi,\eta) = \mathrm{D}\mathrm{E}[f(\xi,\eta)|\xi] + \mathrm{E}\mathrm{D}[f(\xi,\eta)|\xi] \ . \tag{3.47}$$

As an example, let us consider a randomized algorithm of solving the Dirichlet problem for a fourth-order equation:

$$(\Delta + c_1)(\Delta + c_2)u = -g, \quad u\big|_\Gamma = \psi_0, \quad (\Delta + c_2)u\big|_\Gamma = \psi_1.$$

This problem can be reduced to the following system of two Dirichlet problems for the Helmholtz equations:

$$\begin{aligned} &\Delta v + c_1 v = -g, \quad v\big|_\Gamma = \psi_1 \ , \\ &\Delta u + c_2 u = v, \quad u\big|_\Gamma = \psi_0 \ . \end{aligned} \tag{3.48}$$

On this basis, the randomized algorithm is constructed as follows: in solving the main problem (3.48) the values of the function v required for randomized estimation of the integrals over the balls from (3.22) are estimated randomly: each by its own walk trajectory. The unbiasedness of the corresponding estimate is proved using relationship (3.46), while the uniform boundedness of the variance is proved using (3.47) with allowance made for the boundedness of the quantities d_i (see [39]). Thus, one can approximately solve higher-order equations with variable coefficients.

2. The new algorithm can prove to be especially efficient for solving the problems with a random parameter $c(r)$ by the 'double-randomization' method (see section 1.6), which for estimation of the first stochastic moment

$$u_0(r) = \mathrm{E}u(r, c)$$

is expressed by the relationship

$$u_0(r) = \mathrm{EE}(\zeta|c) = \mathrm{EE}(\zeta|\{r_i\}) = \mathrm{E}\zeta.$$

Here ζ is the unbiased (for fixed c) Monte Carlo estimate. The last-mentioned expectation is 'double' and implies that for an unbiased estimate of the quantity $u_0(r)$ to be constructed it is sufficient to construct a single trajectory $\{r_i\}$ for the chosen value of c (see Section 1.6 for the ways of optimizing such a method). However, when using the new algorithm, it is quite feasible to calculate the quantity $\mathrm{E}(\zeta|\{r_i\})$ directly (for a fixed trajectory). For example, for a stochastic scalar parameter c this amounts to calculating the moments of the form of $\mathrm{E}(c^{m_n})$. In the general case for a variable parameter $c(r)$ and higher-order moments it is necessary to calculate mixed power moments (see Section 4.4). This is especially simple for a bounded-correlation model in which the quantities $c(r)$ and $c(r')$ are independent, if $|r - r'| > \rho$; otherwise they coincide.

3. For the Poisson equation with homogeneous boundary conditions, i.e. at $c \equiv 0$ and $\psi = 0$ to construct an unbiased walk on spheres estimate with a finite variance turns out to be possible. To this end, a termination of the chain with a probability of $1 - q$ is simulated in each chain state, while in the case of 'survival', the weight is multiplied by a factor of $1/q$. The unbiasedness of such an estimate is apparent, at least, for $g \geq 0$.

Consider now the estimate $\zeta_1(r)$ obtained from ζ by the replacement $h \to |h| = h_1$ with $r_0 \equiv r$.

The function $\chi(r) = \mathrm{E}\zeta_1^2(r)$ is known to satisfy the following integral equation [39,40]:

$$\chi = h_1(2u_1 - h_1) + K_p\chi,$$

where $u_1 = Ku_1 + h$ (here $k \geq 0$), while the kernel of the integral operator K_p is obtained from the initial kernel $k(r, r')$ by multiplying by a factor of $k(r, r')/p(r'r)$. As we have $p(r', r) \equiv qk(r, r')$ for the unbiased estimate of the solution to the Poisson equation obtained by the walk on spheres technique, then $\mathrm{E}\xi^2 < C_1\chi$, where

$$\chi = q^{-1}K\chi + |h|$$

and the integral operator K corresponds to the Poisson equation, i.e. it implies the averaging over the sphere.

Further, since $\zeta \leq C_2\Sigma d_i^2$, we can construct a chain, of majorization: 'convex domain' → 'plane layer' → 'plane boundary layer' → i.e. we turn to the densities of the centers of spheres of different multiplicities from Section 3.1 with allowance made for the corresponding weights, since

$$\mathrm{E}\Big(\sum d_i^2\Big) = \sum_i q^{-i} \int f_i(x)dx, \tag{3.49}$$

where $f_i(x)$ is the density of the centres of the spheres with a multiplicity of i. From the above it is clear that in (3.49) it is sufficient to consider an integral over a finite interval near zero, for example, over the interval (0,1). Thus, we arrive at a consideration of series (3.6) in which the terms have been multiplied by the corresponding powers of the weight q^{-1}. At the author's suggestion G.A. Syrlybaeva, his post-graduate, proved by means of complex majorization [54] that such a series converged if

$$a > \frac{e \ln 2}{2} \gtrsim 0.942 . \tag{3.50}$$

Thus, when condition (3.50) is fulfilled the unbiased estimate of the solution to the Poisson equation with the zero boundary condition has a finite variance, at least, for a convex domain.

It is interesting to note that the constant in (3.50) turned out to be unimprovable. The point is that the equation

$$f(x) = q^{-1} \int_{x/2}^{\infty} \frac{f(x')}{2x'} dx'$$

has a solution x^{-a} with $q^{-1} = a2^{1-a}$ and $q > \frac{1}{2} e \ln 2$.

Note that so far we failed to prove the possibility of constructing the unbiased estimate for $c > 0$, although this possibility is quite real for sufficiently small values of c. For $c < 0$ such a possibility is evident.

4. The Taylor series considered in Section 3.4, evidently, converges only for the values $c < c^* = c_1^*$, where $\{-c_i^*\}$ are eigenvalues of the Laplacian operator for the domain D. the situation can be improved using the following transformation of problem (3.1):

$$(\Delta + c)v = -(c - c^*)g, \quad v\big|_{\Gamma} = (c - c^*)\psi$$

with both $u(r,c) = v(r,c)/(c - c^*)$ and the Taylor series converging at the point $c_0 < c^*$ for the function $v(r,c)$ when $c < c_2^*$. This series is easy to construct by calculating the derivatives $v^{(k)}(r, c_0)$ with the aid of the algorithms from Section 3.4. Such a method corresponds to analytical continuation of the resolvent by means of eliminating the pole, which method has been applied in [47] to case $g = 0$. Similarly we can take into account other values c_i^*, $i > 1$, and construct efficient expansions for a piecewise constant parameter $c(r)$.

5. The new algorithm can be easily extended to the case of an arbitrary dimension, as has been constructed using the well-known Green function for a ball.

6. The walk on spheres and balls algorithm can be efficiently implemented on multiprocesser systems. First, different trajectories can be 'loaded' into different processors; second, it is possible to calculate the distances to different boundary elements while calculating d_j.

Chapter 4

Solution of metaharmonic equations and elliptic systems

4.1. SOLUTION OF METAHARMONIC EQUATIONS BY CALCULATING THE PARAMETRIC DERIVATIVES

In the Section the required Monte Carlo estimates are constructed and analysed as the estimates of special parametric derivatives on the basis of the following well-known relation between boundary-value problems for metaharmonic equations of different orders.

In a domain $D \subset R^3$ with a boundary Γ, consider the following Dirichlet problem for the Helmholtz equation:

$$\Delta u + cu = -g, \quad u\big|_{\Gamma} = \psi, \tag{4.1}$$

where $c < c^*$ and $-c^*$ is the first eigenvalue of the Laplacian operator in the domain D. The regularity conditions that provide the existence and uniqueness of the solution to problem (4.1) are assumed to be satisfied. These conditions also ensure the probabilistic representation of the solution and its integral representation in terms of the Green function for a ball (see Chapter 3). In particular, it is assumed that $|\psi|, |g| \leq \text{const} < +\infty$.

Straightforward differentiation of problem (4.1) with respect to the parameter c shows that the derivative

$$v_p = u^{(p)} = \frac{\partial^P u}{\partial c^P}$$

is a solution of the following Rykie problem:

$$\begin{aligned} &(\Delta + c)^{p+1} v_p = p!(-1)^{p+1} g \\ &(\Delta + c)^k v_p\big|_{\Gamma} = \frac{(-1)^k p!}{(p-k)!}\psi^{(p-k)}, \quad k = 0, ..., p. \end{aligned} \tag{4.2}$$

Note that relationship (4.2) at $g \equiv 0$ has been in fact used in [47] to construct vector and scalar Monte Carlo algorithms, with the parametric derivatives in the scalar algorithms estimated by using special contour integrals. In Section 3.3 vector 'walk-on-spheres' algorithms have been constructed and justified for calculating the functions v_p at $c \leq 0$. Here we will construct efficient 'walk-on-spheres' algorithms for direct calculation of the parametric derivatives at $c < c^*$, and they will be justified by using the vector representation (see Section 2.3).

To solve problem (4.1), consider the process of 'walk-on-spheres':

$$\{r_n\}, \quad n = 0, 1, ..., N, \quad r_N \in \Gamma_\varepsilon .$$

We have

$$u(r_0) \approx u_\varepsilon(r_0) = \mathrm{E}\eta_\varepsilon^{(0)} ,$$

where

$$\eta_\varepsilon^{(0)} = \sum_{i=1}^{N} \left[\prod_{j=0}^{i-1} s(c, d_j)\right] \int_{D(r_i)} G(\rho; c, d_i) g(\rho) \mathrm{d}\rho + \left[\prod_{j=0}^{N-1} s(c, d_j)\right] \psi(P^*). \tag{4.3}$$

Here P^* is the point on the boundary, which is closest to r_N, $d_j = d(r_j)$ is the distance from point r_j to the boundary, $D(r_i)$ is the ball of radius d_i centred at the point r_i, $G(\rho; c, d)$ is the Green function for a ball, and

$$s(c, d) = \begin{cases} d\sqrt{c}/\sin(d\sqrt{c}), & c \geq 0 \\ d\sqrt{|c|}/\mathrm{sh}(d\sqrt{|c|}), & c \leq 0. \end{cases}$$

Now in contrast to Section 3.2 it is reasonable to randomize the calculation of the integral in (4.3) on the basis of the representation

$$\begin{aligned} \int_{D(r)} G(\rho; c, d_i) g(\rho) \mathrm{d}\rho &= \frac{1}{4\pi} \int_\Omega \mathrm{d}\omega \int_0^d x\left(1 - \frac{x}{d}\right) \frac{s(c, d)}{s(c, d - x)} g(x, \omega) \mathrm{d}x \\ &= \frac{d^2}{6} \mathrm{E}\left\{ \frac{s(c, d)}{s(c, d - \xi)} g(\xi, \omega) \right\}, \end{aligned} \tag{4.4}$$

where ξ is the random quantity distributed over the interval $(0, d)$ according to the density

$$f(x) = \frac{6x(1 - x/d)}{d^2}$$

and ω is an isotropic direction. Thus, instead of (2.1), we may use the estimate

$$\eta_\varepsilon^{(0)} = \sum_{i=1}^{N} \left[\prod_{j=0}^{i-1} s(c, d_j)\right] \frac{d_j^2}{6} \cdot \frac{s(c, d_i)}{s(c, d_i - \xi_i)} g(\xi_i, \omega_i) + \left[\prod_{j=0}^{N-1} s(c, d_j)\right] \psi(P^*). \tag{4.5}$$

Now let us proceed to constructing algorithms for solving the Rykie problem (4.2), i.e. algorithms for calculating parametric derivatives by Monte Carlo method. It is seen from (4.5) that it is sufficient to construct the algorithms for calculating derivatives $Q_{|c|}^{(m)}$, where

$$Q = \prod_{j=0}^{i-1} s(c, d_j).$$

Introduce the notation $a = \sqrt{|c|}$. Differentiating Q with respect to a with the use of the logarithmic derivative

$$Q'_a = Q(\ln Q)'_a$$

at $c \neq 0$ it is easy to obtain the following recurrent relation:

$$Q_a^{(m)} = \sum_{k=1}^{m} u_k Q_a^{(m-k)} \frac{(m-1)!}{(k-1)!(m-k)!}, \quad Q_a^{(0)} = Q,$$

where

$$u_k = (\ln Q)_a^{(k)} = \sum_{j=0}^{i-1} [\ln s(c,d_j)]_a^{(k)} = \sum_{j=0}^{i-1} u_k^{(j)}$$

$$Q_{|c|}^{(1)} = \frac{Q_a^{(1)}}{2a}, \quad Q_{|c|}^{(m)} = \frac{1}{(2a)^m} \sum_{k=1}^{m} a_k^{(m)} a^{k-m} Q^{(k)}, \quad m > 1$$

$$a_1^{(m)} = (-1)^{m-1}(2m-3)!!, \quad a_m^{(m)} = 1, \quad a_k^{(m)} = a_k^{m-1} - (2m-2-k)a_k^{(m-1)}.$$

The quantity $u_k^{(j)}$ is calculated for the sphere with the number j by using the following easy-to-derive formulae:

$$u_k^{(j)} = (k-1)!(-1)^{k-1}a^{-k} + d_j^k f_k, \quad k = 1, 2, \ldots$$

$$f_1 = \begin{cases} -\cos(d_j a)/\sin(d_j a), & c > 0 \\ -\mathrm{ch}(d_j a)/\mathrm{sh}(d_j a), & c < 0 \end{cases}$$

$$\begin{aligned} f_2 &= f_1^2 + \mathrm{sign}(c) \\ f_3 &= 2 f_1 f_2 \\ f_4 &- 2(f_2^2 + f_1 f_3) \\ f_5 &= 2(3 f_2 f_3 + f_1 f_4) \\ f_6 &= 2(3 f_3^2 + 4 f_2 f_4 + f_1 f_5) \\ f_7 &= 2(10 f_3 f_4 + 5 f_2 f_5 + f_1 f_6) \\ f_8 &= 2(10 f_4^2 + 15 f_3 f_5 + 6 f_2 f_6 + f_1 f_7). \end{aligned}$$

At $c = 0$ we have

$$Q_c^{(m)} = \sum_{k=1}^{m} b_k u_k Q_c^{(m-k)} \frac{(m-1)!}{(k-1)!(m-k)!}, \quad Q_c^{(0)} = Q,$$

where

$$b_k u_k = \sum_{j=0}^{i-1} \lim_{c \to +0} [\ln s(c,d_j)]_c^{(k)} = b_k \sum_{j=0}^{i-1} u_k^{(j)}.$$

In this case

$$u_k^{(j)} = d_j^{2k}$$

while the constant coefficients b_k can be obtained by an evident recursion from the representation

$$\begin{aligned} \ln\left[\frac{d\sqrt{c}}{\sin(d\sqrt{c})}\right] &= -\ln\left(\sum_{k=0}^{\infty} \frac{(-1)^k d^{2k}}{(2k+1)!} c^k\right) \\ &= \sum_{p=1}^{\infty} \frac{(-1)^p}{p} \left(\sum_{k=1}^{\infty} \frac{(-1)^k d^{2k}}{(2k+1)!} c^k\right)^p = \sum_{k=1}^{\infty} \frac{d^{2k}}{k!} b_k c^k \end{aligned}$$

which holds for sufficiently small values of c. The first ten coefficients $\{b_k\}$ are given below:

$b_1 = 1.666666667 \cdot 10^{-1}$, $b_6 = 1.298642568 \cdot 10^{-4}$
$b_2 = 1.111111111 \cdot 10^{-2}$, $b_7 = 7.893341227 \cdot 10^{-5}$
$b_3 = 2.116402116 \cdot 10^{-3}$, $b_8 = 5.598081415 \cdot 10^{-5}$
$b_4 = 6.349206349 \cdot 10^{-4}$, $b_9 = 4.537581858 \cdot 10^{-5}$
$b_5 = 2.565335899 \cdot 10^{-4}$, $b_{10} = 4.137766635 \cdot 10^{-5}$.

Now let us discuss the unbiasedness of the estimates of parametric derivatives, i.e. the elements of the solution to the Rykie problem 4.2, and the finiteness of their variances.

Theorem 4.1. *If $c < c^*$ then*

$$\mathrm{E}\eta_\varepsilon^{(m)} = u_\varepsilon^{(m)}(r_0) \quad \forall r_0 \in D.$$

Proof. It is sufficient to consider $c \geq 0$, since for $c < 0$ the theorem is proved similarly. Note that in Section 3.3 the proof for $c < 0$ is given on the basis of even narrower assertion that parametric derivatives for integral equations are unbiased. For $c > 0$ we have

$$s^{(m)}(c, d) = \left(\frac{\sqrt{t}}{\sin\sqrt{t}}\right)_t^{(m)} d^{2m}, \quad m = 1, 2, \ldots$$

and $t = d^2 c < t_0 < \pi^2$. The derivatives of the function $\sqrt{t}/\sin\sqrt{t}$ are bounded uniformly with respect to t, since for sufficiently small t the following equality holds:

$$\frac{\sqrt{t}}{\sin\sqrt{t}} = \left[\sum_{k=0}^{\infty} \frac{(-t)^k}{(2k+1)!}\right]^{-1} = \sum_{i=0}^{\infty}\left[-\sum_{k=0}^{\infty} \frac{(-t)^k}{(2k+1)!}\right]^i. \tag{4.6}$$

We may argue that

$$s^{(m)}(c, d) \leq C_s d^2, \quad m = 1, \ldots, M.$$

Similarly, on the basis of (4.4) and with allowance made for the relation $|g| \leq C_g$, we obtain

$$h^{(m)}(r) \leq \chi(r) = \begin{cases} Cd^2, & r \in D/\Gamma_\varepsilon \\ C, & r \in \Gamma_\varepsilon . \end{cases} \tag{4.7}$$

The same estimate is directly obtained for $c = 0$ on the basis of (4.6). Provided $c < c^*$, the conditions of Theorem 2.6 are fulfilled for the function $\chi(r)$ defined in (4.7). Thus, Theorem 4.1 is proved. □

Theorem 4.2. *If $c < c^*/2$, then*

$$\mathrm{D}\eta_\varepsilon^{(m)} < +\infty \quad \forall r_0 \in D.$$

The proof directly follows from the fact that in this case the function $\chi(r)$ defined in (4.7) satisfies the conditions of Theorem 2.7. □

Note that the ε-biasedness of the estimate $\eta_\varepsilon^{(m)}$, i.e. the relation

$$|u_\varepsilon^{(m)} - u^{(m)}| \leq \text{const} \cdot \varepsilon$$

is justified here in just the same way as in Section 3.4.

It is noteworthy that Sabelfeld [47], on the basis of the mean-value theorem applied to the Rykie problem for $g \equiv 0$, presented a vector version of the estimates constructed in the form of Monte Carlo estimates for the Neumann series associated with 'walk on spheres'. However, he has not finished the justification of the algorithm, namely he has not proved that the specified Neumann series gives a solution to the Rykie problem under study. This could apparently be done by using the probabilistic representation of the solution to the Rykie problem, as it was done for the Helmholtz equation. However, the method we present here is simpler, more natural and universal.

Moreover, the algorithm presented in this paper is scalar and more economic than in [47], since it does not involve matrix operations or estimations of contour integrals by calculating u_λ for the set of values of λ.

The calculations of parametric derivatives for the problem

$$\Delta u + cu = 0, \quad u\big|_\Gamma = 1$$

realize the iterations of the resolvent operator $[\Delta + c]^{-1}$ under homogeneous boundary conditions. Indeed for $u^{(1)}$ we have

$$\Delta u^{(1)} + cu^{(1)} = -u, \quad u^{(1)}\big|_\Gamma = 0.$$

By induction, we then obtain

$$\Delta u^{(m)} + cu^{(m)} = -mu^{(m-1)}, \quad u^{(m)}\big|_\Gamma = 0, \quad m = 1, 2, \dots .$$

Hence, in this case the following relation holds:

$$\frac{mu^{(m-1)}(r_0)}{u^{(m)}(r_0)} \underset{m\to\infty}{\longrightarrow} c^* - c \quad \forall r_0 \in D, \tag{4.8}$$

where $-c^*$ is the first eigenvalue of the Laplacian operator for the domain D.

It is of interest that parametric derivatives also can be used to estimate the eigenvalues of the Helmholtz operator with a variable parameter $c(r)$. Consider

$$\Delta u + [c + c(r)]u = 0, \quad u\big|_\Gamma = 1$$

as the original equation. In this case relationship (4.8) also holds with the same proviso that the first eigenvalue of the Helmholtz operator $\Delta + c(r)$ is considered as $-c^*$. To estimate the corresponding parametric derivatives under certain conditions we also can use the algorithm of 'walk on spheres and balls' which have been developed in Section 3.4. If the poles of the resolvent are separated out, it also can be extended by using the parametric derivatives to construct the appropriate estimates.

By way of example, relationship (4.8) has been numerically implemented for the central points of a sphere S of radius $d_0 = 3^{-1/2}$ and a cube K with edge $h = 1$. In both versions $c^* = 3\pi^2 \approx 29.60881321$. The calculations were performed for three values of the parameter c: 0, 14, and -5. The derivatives $u_S^{(m)}$ with respect to c were exactly calculated for the sphere at $c = 14$ and -5 by the foregoing formulae at $i = 1$, while for $c = 0$ we have

$$u_S^{(m)} = m! a'_m d_0^{2m} = m! a'_m 3^{-m},$$

where a'_m are the coefficients to be given at the beginning of the next section. The estimates of the derivatives $\tilde{u}_K^{(m)}$ for the cube were obtained by using the above-described 'walk-on-spheres' algorithm with the sample size of $N = 25000$ and $\varepsilon = 0.001$ for each version. The computing time for $c \neq 0$ was about two times longer than for $c = 0$.

The results are summarized in Tables 4.1, 4.2 and 4.3, where $\tilde{\sigma}_K^{(m)}$ are the estimates of the root-mean-square error for the cube, while $c_S^{*(m)}$ and $\tilde{c}_K^{*(m)}$ are the estimates of c^* by (4.8). Striking is a monotonic excessive growth of the estimates $c_K^{*(m)}$ for $c = 14$, which is due to a trend to underestimate the highest derivatives by Monte Carlo method at $c \gg 0$. This was observed in different calculation performed in the framework of this research and Section 3.4. Thus, it is reasonable to estimate c^* with $c = 0$.

Table 4.1. Calculation with $c = 0$

m	$u_S^{(m)}$	$c_S^{*(m)}$	$\tilde{u}_K^{(m)} \pm \tilde{\sigma}_K^{(m)}$	$c_K^{*(m)}$
1	$0.555556 \cdot 10^{-1}$	25.7143	$(0.5604 \pm 0.0014) \cdot 10^{-1}$	25.55
2	$0.432099 \cdot 10^{-2}$	28.4516	$(0.4387 \pm 0.0029) \cdot 10^{-2}$	28.38
3	$0.455614 \cdot 10^{-3}$	29.2913	$(0.4638 \pm 0.0061) \cdot 10^{-3}$	29.25
4	$0.622183 \cdot 10^{-4}$	29.5256	$(0.6342 \pm 0.0155) \cdot 10^{-4}$	29.47
5	$0.105303 \cdot 10^{-4}$	29.5875	$(0.1076 \pm 0.0047) \cdot 10^{-4}$	29.50
6	$0.021366 \cdot 10^{-4}$	29.6034	$(0.2189 \pm 0.0160) \cdot 10^{-5}$	29.54
7	$0.005052 \cdot 10^{-4}$	29.6075	$(0.5187 \pm 0.0589) \cdot 10^{-6}$	29.64
8	$0.001365 \cdot 10^{-4}$	–	$(0.1400 \pm 0.0232) \cdot 10^{-6}$	–

Table 4.2. Calculation with $c = 14$

m	$u_S^{(m)}$	$c_S^{*(m)}$	$\tilde{u}_K^{(m)} \pm \tilde{\sigma}_K^{(m)}$	$c_K^{*(m)}$
0	2.598805	–	2.620 ± 0.015	–
1	0.226902	28.744	0.2304 ± 0.0048	28.72
2	$0.307799 \cdot 10^{-1}$	29.456	$(0.3131 \pm 0.0174) \cdot 10^{-1}$	29.57
3	$0.597454 \cdot 10^{-2}$	29.584	$(0.6031 \pm 0.0692) \cdot 10^{-2}$	29.98
4	$0.153351 \cdot 10^{-2}$	29.605	$(0.1510 \pm 0.0293) \cdot 10^{-2}$	30.46
5	$0.491353 \cdot 10^{-3}$	29.608	$(0.4586 \pm 0.1297) \cdot 10^{-3}$	31.10
6	$0.188882 \cdot 10^{-3}$	29.609	$(0.1609 \pm 0.0596) \cdot 10^{-3}$	31.86
7	$0.847075 \cdot 10^{-4}$	29.609	$(0.6307 \pm 0.2837) \cdot 10^{-4}$	32.70
8	$0.434153 \cdot 10^{-4}$	–	$(0.2698 \pm 0.1398) \cdot 10^{-4}$	–

Additionally note that, in principle, it is possible here to use the well-known Cauchy-Hadamard dormula for the convergence radius of a power series on the basis of the expression:

$$u(r,c) = \sum_{m=0}^{\infty} \frac{u^{(m)}(r,c_0)}{m!}(c-c_0) = \sum_{m=0}^{\infty} a_m(c-c_0).$$

Here the Cauchy-Hadamard formula can be represented in the form:

$$c^* - c_0 = \lim_{m\to\infty} \left(1/\sqrt[m]{a_m}\right).$$

Computations show that the corresponding convergence in this case is comparatively slow. For instance, when $c_0 = 0$ there was obtained the following sequence of the values $1/\sqrt[m]{a_m}$: 18, 21.5, 23.6, 24.9, 25.8, 26.4, 26.8, 27.2 instead of the values $\{c_S^{*(m)}\}$ from Table 4.1.

Table 4.3. Calculation with $c = -5$

m	$u_S^{(m)}$	$c_S^{*(m)}$	$\tilde{u}_K^{(m)} \pm \tilde{\sigma}_K^{(m)}$	$c_K^{*(m)}$
0	0.768128	–	0.76637 ± 0.00046	–
1	$0.385775 \cdot 10^{-1}$	24.159	$(0.3882 \pm 0.0006) \cdot 10^{-1}$	23.94
2	$0.264605 \cdot 10^{-2}$	27.797	$(0.2682 \pm 0.0011) \cdot 10^{-2}$	27.68
3	$0.242040 \cdot 10^{-3}$	29.050	$(0.2462 \pm 0.0020) \cdot 10^{-3}$	29.02
4	$0.284334 \cdot 10^{-4}$	29.444	$(0.2895 \pm 0.0042) \cdot 10^{-4}$	29.42
5	$0.041275 \cdot 10^{-4}$	29.561	$(0.4206 \pm 0.0102) \cdot 10^{-5}$	29.50
6	$0.007165 \cdot 10^{-4}$	29.595	$(0.7315 \pm 0.0289) \cdot 10^{-6}$	29.49
7	$0.001450 \cdot 10^{-4}$	29.605	$(0.1485 \pm 0.0093) \cdot 10^{-6}$	29.49
8	$0.000335 \cdot 10^{-4}$	–	$(0.0344 \pm 0.0033) \cdot 10^{-6}$	–

4.2. SOLVING METAHARMONIC EQUATIONS OF THE FORM: $\Delta^{p+1}u + cu = (-1)^{p+1}g$

First, let us construct a 'walk-on-spheres' algorithm for solving the first boundary-value problem in the domain D with the boundary Γ for the nonhomogeneous polyharmonic equation $\Delta^{p+1} = (-1)^{p+1}g$ on the basis of formula (4.2) at $c \equiv 0$.

Note that the corresponding system (2.13) is randomized here in the form of (2.28), where

$$a_1^{(m-i)} = s^{(m-i)}(0,d) = s^{(m-i)}(c,d)\big|_{c=0}$$

and we may take $c > 0$, i.e.

$$s(c,d) = \frac{d\sqrt{c}}{\sin(d\sqrt{c})}\ .$$

Without loss of generality, we assume that the following relationship holds:

$$d \le d_{\max} \le 1 \tag{4.9}$$

since we can arrive at this relationship by changing properly the scale.

In the case $d_{\max} = 1$ the inequality

$$c^* \leq \pi^2$$

evidently holds by virtue of the fact that the value of c^* for a ball of unit radius equals π^2.

Let us make the following substitutions: consider the functions $\psi_m = \varphi_m/m!$ instead of the functions φ_m in system (2.13), and accordingly, the quantities $\eta^{(m)} = \xi^{(m)}/m!$ instead of the random quantities $\xi^{(m)}$ in system (2.28). Then system (2.28) takes the form:

$$\eta_0^{(m)} = \sum_{i=0}^{m} \frac{s^{(m-i)}(0, d_0)}{(m-i)!}\eta_1^{(i)} + h_0^{(m)}/m!, \quad m = 0, 1, ..., p. \tag{4.10}$$

Note that the quantities

$$q_{m-i} = \frac{s^{(m-i)}(0, d)}{(m-i)!}$$

are the coefficients of the Maclaurin series for the function $s(c, d)$ and can be found recurrently on the basis of the representation:

$$s(c, d) = \sum_{i=0}^{\infty} \left[\sum_{k=1}^{\infty} (-1)^{k-1} \frac{(d^2c)^k}{(2k+1)!} \right]^i = \sum_{k=0}^{\infty} a_k c^k.$$

Clearly, $a_k = a'_k d^{2k}$, where a'_k are the coefficients of the Maclaurin series for the function $s(c, 1)$. The first ten coefficients a'_k are given below:

$$\begin{aligned}
a'_1 &= 1.66666667 \cdot 10^{-1}, & a'_4 &= 2.09986773 \cdot 10^{-4} \\
a'_2 &= 1.94444444 \cdot 10^{-2}, & a'_5 &= 2.13360456 \cdot 10^{-5} \\
a'_3 &= 2.05026455 \cdot 10^{-3}, & a'_6 &= 2.16334744 \cdot 10^{-6} \\
a'_7 &= 2.19232713 \cdot 10^{-7}, & a'_9 &= 2.25076748 \cdot 10^{-9} \\
a'_8 &= 2.22139309 \cdot 10^{-8}, & a'_{10} &= 2.28051077 \cdot 10^{-10}.
\end{aligned}$$

These coefficients are positive, since (see Section 3.2)

$$s(c, d) = \mathrm{E}e^{c\tau}, \quad s^{(k)}(0, d) = \mathrm{E}(\tau^k),$$

where τ is the point in time at which the Wiener process, originating at the centre of a sphere of radius d, leaves this sphere for the first time. In addition, we have

$$\sum_{k=0}^{\infty} a'_k = \frac{1}{\sin(1)} = 1 + \sum_{k=1}^{\infty} a'_k = 1 + 0.188... . \tag{4.11}$$

In this case the probabilistic representation (4.10) of modified system (2.13) is connected with 'walk on spheres' $\{r_n\}_{n=0}^{N}$ and $\eta_0^{(m)} \equiv \eta_{r_0}^{(m)}$ is the m-th component of the corresponding vector estimate. As mentioned, for example, in [36], using additional randomization over the component number one can go from the vector estimate to the scalar estimate which corresponds to the representation of system (2.13) in the form of one integral equation. To make this transition, let us introduce one more integral-valued coordinate m in the phase space of the chain simulated. The value of this coordinate is re-chosen according to the probabilities:

$$P(m_n = i | m_{n-1} = m) = a_{m-i} > 0, \quad i = 0, 1, ..., m-1$$

$$P(m_n = m | m_{n-1} = m) = 1 - \sum_{i=0}^{m-1} a_{m-i} = P_m(d).$$

Each next value of r_n is chosen on the sphere $S(r_{n-1})$ in the usual fashion [13]. Thus, instead of the chain $\{r_n\}$ we construct the chain $\{x_n\}$, and

$$x_n = (r_n, m_n).$$

Additionally, we calculate a random scalar weight by the formulae:

$$Q_0 \equiv 1, \quad Q_n = Q_{n-1} q_n, \quad q_n = \begin{cases} 1, & m_n \neq m_{n-1} \\ 1/P_{m_n}(d), & m_n = m_{n-1}. \end{cases}$$

The corresponding scalar estimate has the form:

$$\xi_0 = \xi_{x_0} = \xi_{(r_0, m_0)} = \sum_{n=0}^{N} \frac{Q_n h^{(m_n)}(r_n)}{m_n!} . \tag{4.12}$$

Evidently, the conditional expectation of this estimate with a fixed value of the chain $\{r_n\}$ equals the appropriate value of the vector estimate $\eta_0^{(m_0)}$. Hence at $m_0 = p$ we have:

$$\mathrm{E}\xi_0 = \psi_p(r_0) = \frac{u^{(p)}(r_0)}{p!} .$$

To take advantage of the statement in Section 3.4 that the variances of the estimates are finite, we modify ξ_0 by calculation $h^{(m)}(r)$ at $r \notin \Gamma_\varepsilon$ only for the transition $(r, m) \to (r', m)$, i.e. only if m is preserved, however with the weight q/P_m. The modified estimate has the form:

$$\xi_{0,1} = \sum_{i=1}^{n_N} \frac{Q_{n_i} h^{(m_{n_i})}(r_{n_i - 1})}{m_{n_i}!} + \frac{Q_N \psi^{(m_N)}(P^*(r_N))}{m_N!} .$$

Here n_i are the numbers of the transitions preserving m. The repeated averaging yields $\mathrm{E}\xi_{0,1} = \xi_0$.

Taking into account that $d_n \leq 1$, we have

$$q_n \leq \frac{1}{1 - \sum_{i=0}^{m-1} d_{n-1}^{2(m-i)} a'_{m-i}} \leq \frac{1}{1 - d_{n-1}^2 \sum_{k=1}^{\infty} a'_k} \leq \frac{1}{1 - 0.189 d_{n-1}^2} . \tag{4.13}$$

Now notice that the following assertion follows from the proof of Theorem 3.4: if

$$q_n \leq \frac{1}{1 - \frac{1}{6} c_0 d_{n-1}^2} \tag{4.14}$$

at $c_0 \leq 0.488 c^*$, then, with allowance made for estimate (4.7), the following relation holds for $h^{(m)}$:

$$\mathrm{D}\xi_{0,1} \le C_D < \infty \,.$$

According to (4.13), we should take here $c_0 = 6 \cdot 0.189$ and then we get $\mathrm{D}\xi_{0,1} < +\infty$, if $c^* > 2.324$.

For subsequent constructions we need to define the weight function in the integral that expresses $h^{(m)}(r)$ at $c = 0$ and $r \notin \Gamma_\varepsilon$.

According to (2.2), we have

$$h^{(m)}(r) = \frac{1}{4\pi}\int_\Omega \mathrm{d}\omega \int_0^d x\Big(1-\frac{x}{d}\Big)\Big[\frac{s(c,d)}{s(c,d-x)}\Big]^{(m)}_{c-0} g(x)\mathrm{d}x \,. \tag{4.15}$$

Denote the m-multiple derivative appearing in (4.15) by $H^{(m)}(0,x)$. Using the representation

$$\frac{s(c,d)}{s(c,d-x)} = \Big\{\sum_{i=0}^\infty \Big[\sum_{k=1}^\infty (-1)^{k-1}\frac{(d^2c)^k}{(2k+1)!}\Big]^i\Big\}\Big[\sum_{k=0}^\infty (-1)^k \frac{(d-x)^{2k}c^k}{(2k+1)!}\Big]$$
$$= \Big(\sum_{k=0}^\infty a'_k d^{2k}c^k\Big)\Big[\sum_{k=0}^\infty (-1)^k\frac{(d-x)^{2k}c^k}{(2k+1)!}\Big] = \sum_{k=0}^\infty \frac{H^{(k)}}{k!}c^k$$

we obtain

$$\frac{H^{(m)}(0,x)}{m!} = \sum_{k=0}^m \frac{(-1)^{m-k}a'_k d^{2k}(d-x)^{2(m-k)}}{[2(m-k)+1]!} \,.$$

Note that we have thus defined the expression for the Green function $G^{(m)}(x;0,d)$ for a ball of radius d with a pole at the centre of the ball for the polyharmonic equation, since

$$4\pi x G^{(m)}(x;0,d) = \Big(1-\frac{x}{d}\Big)\frac{H^{(m)}(0,x)}{m!} \,.$$

In particular, for $m = 1$ this implies that

$$H^{(1)}(0,x) = \frac{d^2-(d-x)^2}{6} \,. \tag{4.16}$$

It is reasonable to randomize the calculation of $h^{(m)}(r)$ here by choosing the random value x according to the density, which is proportional to the function

$$g_m(x) = x\Big(1-\frac{x}{d}\Big)H^{(m)}(0,x) \le C_m x\Big(1-\frac{x}{d}\Big) \,.$$

The last-mentioned inequality allows the method of elimination to be used for this purpose (see, for example, [39]) with the preliminary choice of ρ according to the density $6x(1-x/d)d^{-2}$. The weight factor of the random value $g(x)$ is determined by the expression

$$G_m = \int_0^d x\Big(1-\frac{x}{d}\Big)H^{(m)}(0,x)\mathrm{d}x \,.$$

Note that $H^{(m)}(0,x) \ge 0$, since the following relation holds for any nonnegative integrable function g:

$$h^{(m)}(r) = \left[\mathrm{E}\int_0^{\tau} e^{ct} g(\xi(t))dt\right]^{(m)}_{c=0} = \mathrm{E}\int_0^{\tau} t^m g(\xi(t))dt \geq 0,$$

where $\xi(t)$ is the above-defined Wiener process and τ are the point in time at which it leaves the domain D for the first time.

It is also clear that

$$G_m = \mathrm{E}\int_0^{\tau} t^m dt = \frac{1}{m+1}\mathrm{E}\tau^{m+1} = \frac{s^{(m+1)}}{m+1} = m!a'_{m+1}d^{2(m+1)}.$$

For example, at $m = 1$, using (4.16), we obtain

$$g_1(x) = \frac{dx^2}{6}\left(1-\frac{x}{d}\right)\left(2-\frac{x}{d}\right), \quad 0 \leq x \leq d.$$

Passing to the interval (0,1) up to a constant factor we have:

$$g_1(y) = y^2(1-y)(2-y) \leq y(1-y) = p_1(y).$$

The probability of the corresponding elimination equals

$$1 - \frac{\int g_1(y)dy}{\int p_1(y)dy} = 0.3.$$

Note that $G_1 = 7d^4/360$. The following estimate, which result from the inequality $d \leq 1$ and relation (4.11), is helpful in what follows:

$$\frac{G_m}{m!} \leq a'_{m+1}d^2 \leq \left(\sum_{k=2}^{\infty} a'_k\right)d^2 < 0.023d^2, \quad m \geq 1.$$

Now let us proceed to constructing the algorithm for solving the equation

$$\Delta^{p+1}u + cu = (-1)^{p+1}g$$

with variable $c = c(r)$. To this end, similar to Section 3.4, we apply the above-considered randomization once the equation is transformed to the form:

$$\Delta^{p+1}u = -cu + (-1)^{p+1}g\ .$$

Then, instead of (4.10), we have the following system:

$$\eta_0^{(m)} = \sum_{i=0}^{m} \frac{s^{(m-i)}(0,d_0)}{(m-i)!}\eta_1^{(i)} + (-1)^p\frac{c}{m!}G_m\eta_{01}^{(m)} + \frac{G_m}{m!}g(r_1), \tag{4.17}$$

where the random quantity $\eta_{01}^{(m)}$ in the scalar algorithm is constructed on the chain originating at a random point r_1, which is chosen within the ball $D(r_0)$ according to the density $g_m(x)/G_m$. Thus, it is necessary here to simulate a chain of 'walk on spheres and balls': a 'jump' inside a ball while preserving the value of m occurs with the probability $c_1G_m/m!$, the weight being multiplied by $(-1)^pc/c_1$. In the sequel, we assume that $|c|/c_1 \leq 1$. The probabilities of transitions $m_n \to m_{n-1}$ on a sphere are given as it was done previously in solving the equation $\Delta^{p+1}u = (-1)^{p+1}g$, with the difference that the quantity P_m (and hence the weight factor q_n after a 'jump' onto a sphere) is determined by the expression:

$$P(m_n = m \cap r_n \in S(r_{n-1}) | m_{n-1} = m) = 1 - \sum_{i=0}^{m-1} a_{m-i} - \frac{c_1}{m!} G_m \, .$$

Instead of (4.13), we can use here the inequality:

$$q_n \le \begin{cases} \left(1 - \frac{1}{6} c_1 d_{n-1}^2\right)^{-1}, & m_{n-1} = 0 \\ (1 - 0.189 d_{n-1}^2 - 0.023 c_1 d_{n-1}^2)^{-1}, & m_{n-1} > 0. \end{cases}$$

By repeating the reasoning of Lemmas 3.1 and 3.2 and Theorem 3.1 and 3.2, it is easy to obtain the following assertion: the expectation of random quantity (4.12) at $m_0 = p$ is expressed here by the Neumann series, corresponding to representation (4.17), and yields an ε-biased solution of the problem:

$$\Delta^{p+1}u + cu = (-1)^{p+1}g$$

$$\Delta^k u\big|_\Gamma = \frac{(-1)^k}{(p-k)!}\psi^{(p-k)}, \quad k = 0, 1, \ldots, p$$

if inequality (4.14) holds and $c_0 \le 0.6079c^*$. In this case the required condition is fulfilled if $c_1 \le 0.6079c^*$ and $6(0.189 + 0.023c_1) < 0.6079c^*$, since

$$1 - 0.189 d_{n-1}^2 - 0.023 c_1 d_{n-1}^2 = 1 - \frac{1}{6} d_{n-1}^2 (6 \cdot 0.189 + 6 \cdot 0.023 c_1) \, .$$

The scalar randomized version of the estimate $\eta_0^{(m)}$ will be denote by η_0.

Now, let us turn from the estimate η_0 to the estimate $\eta_{0,1}$ by calculating $h^{(m)}(r)$ at $r \notin \Gamma_\varepsilon$ only for a transition $(r', m) \to (r, m)$ on a sphere, but with the weight $1/P_m$. With allowance made for the latter equality, Theorem 3.4 shows that the variance of the estimate $\eta_{0,1}$ is uniformly bounded with respect to ε, if the inequalities $c_1 \le 0.488c^*$ and $6(0.189 + 0.023c_1) < 0.488c^*$ hold.

One further comment should e made: according to (4.3) we have

$$h^{(m)}(r_N) = \psi^{(m)}(P_N^*), \quad r_N \in \Gamma_\varepsilon$$

and here it is also assumed that the parametric derivatives of the function $\psi = \psi(r, c)$ are uniformly bounded.

Evidently, within the limits of the algorithm constructed, we can simultaneously estimate the solutions of problems of the form of (4.2) for different values of the parameter $p' < p$.

For this purpose a vector Monte Carlo algorithm (see Section 2.1) can be essentially useful. This algorithm is constructed here on the' walk on spheres and balls' in the following way. With the probability

$$1 - c_1 G_1 d^2 = 1 - C_1 d^2/6$$

a new point is chosen uniformly over the sphere S, while the matrix weight $\tilde{Q}$ is multiplied y the matrix of system (4.10) and divided by the value $1 - c_1 d^2/6$; with the opposite probability $c_1 d^2/6$ a new point is chosen within the ball D, while the matrix weight is multiplied by the diagonal matrix

$$\mathrm{diag}\{(-1)^p \frac{c}{m!} G_m/(c_1 d^2/6)\} .$$

Here

$$(-1)^p \frac{c}{m!} G_m/(c_1 d^2/6) < 1.$$

The corresponding vector estimate has the form:

$$\tilde{\xi}_0 = \sum_{n=0}^{N} \tilde{Q}_n \tilde{h}(r_n), \quad Q_0 = \mathrm{diag}(1, 1, ..., 1).$$

A transition from the homogeneous version to the nonhomogeneous one is performed here similarly to the case of a scalar estimate. By using the matrix norm $\|\cdot\|_M$ the following relation is obtained:

$$\|\tilde{\xi}_0\|_M \leq \sum_{n=0}^{N} \|\tilde{Q}_n\|_M \, \|\tilde{h}(r_n)\|_M .$$

For an elementary weight multiplier we have here the estimate

$$\|q\|_M \leq \frac{1 + 0.189 d^2}{1 + c_1 d^2/6} \leq \frac{1}{1 - (c_1 + 0.189 \cdot 6) d^2/6} .$$

Therefore a variance of a vector estimate is finite provided the following condition is fulfilled:

$$c_1 + 0.189 \cdot 6 < 0.488 \cdot c^*$$

4.3. TWO-DIMENSIONAL CASE

The following integral representation is valid for the solution to problem (4.1) at $c > 0$ in the two-dimensional case:

$$u(r) = \frac{1}{2\pi d(r) J_0(d\sqrt{c})} \int_{S(r)} u(r(s)) ds + \int_{D(r)} G(\rho; c, r) g(\rho) d\rho . \tag{4.18}$$

Here $J_0(d\sqrt{c})$ is the Bessel function, i.e.

$$J_0(d\sqrt{c}) = \sum_{k=0}^{\infty} \frac{(-1)^k d^{2k} c^k}{d^{2k} (k!)^2} , \quad d = d(r) \tag{4.19}$$

and $G(\rho; c, r)$ is a 'circular' Green function of Helmholtz equation for a circle $D(r)$ with the pole at its centre. The circumference $S(r)$ is the boundary of the circle. The circumference is chosen so that it should be tangent to the boundary Γ, with $d(r)$ being the distance between the point r and the boundary Γ.

On the basis of (4.18), we construct a 'walk-on-circumferences' algorithm similarly to the three-dimensional case (see Section 3.4). The corresponding estimate has the form of (4.3), with the difference that

$$s(c, d) = \frac{1}{J_0(d\sqrt{c})} .$$

Similar to the three-dimensional case, here $u(r_0) \approx u_\varepsilon(r_0) = \mathrm{E}\eta_\varepsilon^{(0)}$, if $c < c^*$.

In the following, we will use the relation:

$$\tilde{J}(d\sqrt{c^*}) = 1 - \frac{d^2c^*}{4} + \frac{(d^2c^*)^2}{64} > J_0(d\sqrt{c^*}) .$$

Lemma 4.1. *If $c < c^*/2$, then*

$$\mathrm{D}\eta_\varepsilon^{(0)} < C_D < +\infty \quad \forall \varepsilon > 0 .$$

Proof. For $g \equiv 0$ that statement of the lemma follows from the relation:

$$J^2(d\sqrt{c^*/2}) \geq (1 - d^2c^*/8)^2 = \tilde{J}(d\sqrt{c^*}).$$

The transition to the nonhomogeneous case then performed by using a special representation of the weight Q_n, which is given in Section 3.2. Lemma 4.1 is proved. □

Now let us consider the algorithm of 'walk on circumferences and circles' for solving problem (4.1) at $c = c(r)$. Passing on to the equation $\Delta u = -cu - g$ in (4.1), we obtain an integral representation, which in $D\backslash\Gamma_\varepsilon$ has the form:

$$u(r) = \frac{1}{2\pi d(r)} \int_{S(r)} u(r(s))\mathrm{d}s + \int_{D(r)} c(r')G(\rho;0,r)u(\rho)\mathrm{d}\rho + \int_{D(r)} G(\rho;0,r)g(\rho)\mathrm{d}\rho, \quad r \in D\backslash\Gamma_\varepsilon \tag{4.20}$$

and is an identity in Γ_ε:

$$u(r) = u(r), \quad r \in \Gamma_\varepsilon .$$

It is known that

$$G(\rho;0,r) = \frac{\ln(d/|\rho - r|)}{2\pi}, \quad \int_{D(r)} G(\rho;0,r)\mathrm{d}\rho = \frac{1}{4}d^2.$$

The randomization of presentation (4.20), necessary for constructing the 'walk-on-circumferences' estimate is made upon passing in the right side of (4.20) to the following expression:

$$\begin{aligned}&\left(1 - \frac{1}{4}c_0d^2\right)\frac{(1 - \frac{1}{4}c_0d^2)^{-1}}{2\pi d}\int_{S(r)} u(r(s))\mathrm{d}s\\&+\frac{1}{4}c_0d^2\int_{D(r)}\frac{c(\rho)\cdot 4}{c_0d^2}G(\rho;0,r)u(\rho)\mathrm{d}\rho + h(r),\end{aligned} \tag{4.21}$$

where c_0 should obey the inequality:

$$\frac{1}{4}c_0d_{\max} < 1 .$$

According to (4.21), the Markov chain of 'walk on circumferences and circles' is constructed as follows: with the probability $1 - c_0d^2(r_0)/4$ a new point r_{i+1} is selected uniformly on the circumference $S(r_i)$, while the weight Q is multiplied by

$$q(r_i, c_0) = \frac{1}{1 - \frac{1}{4}c_0d^2(r_i)}$$

and with the probability $c_0d^2(r_i)/4$ the point r_{i+1} is selected within the circle $D(r_i)$ according to the density $4d^{-2}(r_i)G(\rho;c,r)$, while the weight is multiplied by c/c_0. When the chain gets inside Γ_ε, it terminates and the quantity $Q_N\psi(P)$ is added to the sum. The resulting random estimate has the form:

$$\xi_\varepsilon = \sum_{n=0}^{N} Q_n h(r_n),$$

where

$$Q_n = \left[\prod_{i=1}^{m_n} \frac{c(r_{k_i})}{c_0}\right]\left[\prod_{i=1}^{n-m_n} q(r_{k_i}, c_0)\right]$$

while the function $h(r)$ is calculated either exactly or in a randomized manner (see below). Here $\{r_{t_i}\}$, $i = 1, ..., n - m_n$, are the points selected on the circumferences, $\{r_{k_i}\}$, $i = 1, ..., m_n$, are the points selected within the circles.

Now we justify asymptotic unbiasedness of the quantity ξ_ε as the estimate of the solution to the initial problem.

Hereafter the symbol x^* stands for the least root of the equation $J_0(x) = 0$. It is known that $x^* \approx 2.405$. Clearly, the inequality $r\sqrt{c^*} < x^*$ has to hold.

Lemma 4.2. *Let $D \subset R^2$ and let the first derivatives of the solution to problem (4.1) be uniformly bounded in D, $c(r)| < c_0$, and*

$$\frac{c_0}{c^*} < 1 - \frac{x^{*2}}{16} \lesssim 0.638. \tag{4.22}$$

Then

$$\mathrm{E}\xi_\varepsilon = u_\varepsilon, \quad |u(r) - u_\varepsilon(r)| \leq \mathrm{const}\cdot\varepsilon, \quad r \in D, \quad \varepsilon > 0.$$

Proof. It is easy to verify that at $r\sqrt{c^*} < x^*$ the inequality $\tilde{J}(d\sqrt{c^*}) < 1 - c_0 d^2/4$ follows from (4.22), and hence

$$|q(r, c_0)| < \frac{1}{J_0(d\sqrt{c^*})}.$$

Thus, the quantity ξ_ε is termwise majorized by the standard estimate ζ on the 'walk on circumferences' (it does not have to be maximum) with a certain value $c < c^*$ at $\varepsilon = 0$. The Neumann series corresponding to the estimate ζ converges, since ζ can be obtained by the partial averaging of the probabilistic representation of the solution. The proof is completed by literal repetition of the reasoning from Lemma 3.2 and Theorems 3.1 and 3.2. □

As in three-dimensional case to extend the conditions for the finiteness of the variance, we modify the estimate in the following way. Let us calculate the function $h(r)$ at $r \notin \Gamma_\varepsilon$ only for the step $r \to r'$ onto a sphere, but with the weight $q(r, c_0)$, i.e. let us consider the estimate

$$\xi_{\varepsilon,1} = \sum_{n=0}^{N} Q_n h_1(r_n) = \sum_{i=1}^{N-m_N} Q_{t_i} h(r_{t_i-1}) + Q_N \psi(P^*(r_N))$$

instead of the estimate ξ_ε. The repeated averaging yields $\mathrm{E}\xi_{\varepsilon,1} = \mathrm{E}\xi_\varepsilon$.

Lemma 4.3. *If the conditions $|c(r)| \leq c_0$ and $c_0 < c^*/2$ are met then*

$$\mathrm{D}\xi_{\varepsilon,1} < C_D < +\infty \quad \forall\varepsilon > 0.$$

Proof. If $c_0 = c^*/2$ then

$$\tilde{J}(d\sqrt{c^*}) = \left(1 - \frac{1}{8}d^2c^*\right)^2 = \left(1 - \frac{1}{4}d^2c_0\right)^2 > J_0(d\sqrt{c^*}).$$

Hence, if $c_0 < c^*/2$ then

$$q^2(r, c_0) < \frac{1}{J_0(d\sqrt{c_1})}$$

for certain $c_1 < c^*$, i.e. the statement of the lemma is true for $g \equiv 0$. The transition to the nonhomogeneous case is made in just the same way as in proving Theorem 3.4: on the basis of the majorization with the use of $c \equiv c_0 < c^*/2$ and $h(r) = \text{const} \cdot d^2$. Lemma 4.3 is proved. □

When proceeding to the Monte Carlo algorithm for the equation

$$\Delta^{p+1}u = -cu + (-1)^{p+1}g$$

in the two-dimensional case, it should first of all be noted that representation (4.17) holds for any dimension, and hence the algorithm of 'walk on circumferences and circles' is well defined, together with the corresponding values of the constants G_m and the densities $g_m(x)$. As previously, $G_m = m!a'_{m+1}d^{2(m+1)}$, but the quantities a'_m are determined from the relation

$$\frac{1}{J_0(d\sqrt{c_1})} = \sum_{i=0}^{\infty}\left[\sum_{k=1}^{\infty}(-1)^{k-1}\frac{(d^2c)^k}{2^{2k}(k!)^2}\right]^i = \sum_{k=0}^{\infty} a_k c^k,$$

where $a_k = a'_k d^{2k}$. Here

$$\sum_{k=0}^{\infty} a'_k = \frac{1}{J_0(1)} = 1 + \sum_{k=1}^{\infty} a'_k = 1 + 0.30685...$$

and

$$\frac{G_m}{m!} \leq a'_{m+1}d^2 \leq \left(\sum_{k=2}^{\infty} a'_k\right)d^2 < 0.057d^2, \quad m > 1.$$

Repeating literally the reasoning of Section 3, we arrive at analogous statement about the ε-biasedness of the estimate η_0, and about uniform boundedness of the variance with the following change of constants: $0.6079 \to 0.638$, $0.488 \to 0.5$, $0.189 \to 0.307$, $0.023 \to 0.057$, $6 \to 4$.

Then the Green function $G(\rho; c, r)$ is easily obtained from the fundamental solution of the Helmholtz equation [50]:

$$G(\rho; c, r) = \frac{1}{4}\left[-N_0(z\sqrt{c}) + \frac{N_0(d\sqrt{c})J_0(z\sqrt{c})}{J_0(d\sqrt{c})}\right], \tag{4.23}$$

where $z = |\rho - r|$ and $N_0(z)$ is the Neumann function:

$$N_0(z) = \frac{2}{\pi}J_0(z)\ln\frac{z}{2} - \frac{2}{\pi}\sum_{k=1}^{\infty}\frac{(-1)^k}{(k!)^2}\left(\frac{z}{2}\right)^{2k}\left(\frac{1}{k} + \frac{1}{k-1} + ... + 1\right).$$

Using the asymptotic expressions when $c \to 0$:

$$J_0(z\sqrt{c}) \sim 1, \quad N_0(z\sqrt{c}) \sim \frac{2}{\pi}\ln\frac{z\sqrt{c}}{2}$$

$$J_0'(z\sqrt{c}) \sim \frac{z^2}{4}, \quad N_0'(z\sqrt{c}) \sim \frac{r^2}{4}\ln\frac{z\sqrt{c}}{2} + \frac{1}{2\sqrt{c}} + \frac{z^2}{4}$$

we obtain from (4.23) the well-known formulae (see, for example [47]):

$$G(\rho;0,r) = \frac{1}{2\pi}\ln\frac{d}{z} = \frac{1}{2\pi}\cdot\frac{d^2}{4}g_0(z), \quad z = |\rho - r|$$

$$G^{(1)}(\rho;0,r) = \frac{1}{8\pi}\left(d^2 - z^2 - z^2\ln\frac{d}{z}\right) = \frac{1}{2\pi}\cdot\frac{3d^4}{64}g_1(z).$$

The functions $zg_0(z)$ and $zg_1(z)$ are the distribution densities that after the variable change $x = z/d$ take on the form:

$$p_0(x) = -4x\ln x \le 4e^{-1}$$

$$p_1(x) = \frac{16}{3}x(1 - x^2 + x^2\ln x) \le 1.539, \quad 0 < x < 1 .$$

The simplest version of the method of elimination (see, for example, [39]) is fairly efficient for simulating random quantities with these densities. The corresponding elimination probabilities are estimated by the values 0.321 and 0.531.

Thus, we have given a complete definition of the new Monte Carlo algorithm, for example, for solving the problem

$$\Delta^2 u + c(r)u = g(r), \quad u\big|_\Gamma = \Delta u\big|_\Gamma = 0 \tag{4.24}$$

which is encountered in the theory of elasticity (see, for example, [47]). The algorithm constructed is especially convenient if $c(r)$ and $g(r)$ are random functions and the problem is solved by the double-randomization method [40]. The appropriate estimates of the solution moments can be preliminary averaged over the distributions $c(r)$ and $g(r)$, which makes it possible to use only power probability moments of the functions $c(r)$ and $g(r)$. These moments are well defined, for example, for homogeneous Gaussian $c(r)$ and $g(r)$ by their covariance functions. Hence we have constructed an ε-biased estimate of 'walk on circumferences and circles' for the complete solution to problem (4.24) with homogeneous Gaussian functions $c(r)$ and $g(r)$ without their preliminary approximation.

It is interesting to note that, based on the probabilistic representation constructed, we may argue that at $c(r) \equiv 0$ the second probability moments of the random solution to problem (4.24) are well defined by the second moments of the random function $g(r)$.

4.4. CALCULATION OF THE COVARIANCE FUNCTION OF THE SOLUTION TO THE BIHARMONIC EQUATION

The section is devoted to the new Monte Carlo method for the calculation of the covariance function of the solution of the biharmonic equation when its right-hand side is a random field. The comparison of this method with the randomized algorithm of the Monte Carlo method (see Section 1.6) is presented. The numerical results of the solution of the concrete problem are given [55].

Let us consider the problem of a plate oscillation in the bounded domain $G \subset R^2$ under the influence of the random loading field $\sigma = \sigma(\bar{x})$. The solution of the boundary value problem

$$\Delta^2 u(\bar{x}) = \sigma(\bar{x}), \quad u\big|_{\partial G} = \varphi(\bar{x}), \quad \Delta u\big|_{\partial G} = \psi(\bar{x}) \tag{4.25}$$

is also the random field. It is necessary to define it's covariance function $v(\bar{x}, \bar{y}) = \mathrm{E}\{u(\bar{x})u(\bar{y})\}$ at the points $\bar{x}_0$ and $\bar{y}_0$. In the methods existed earlier [47] the product of the solution estimates in two given points for different realizations of the field $\sigma(\bar{x})$ was averaged (so called randomized algorithms of the Monte Carlo method (see Section 1.6)). The method given below enables us to use the covariance function $K(\bar{x}, \bar{y})$ of the field σ directly.

Let us multiply the initial equation by the same equation depending on the other variable $\bar{y} : \Delta^2 u(\bar{y}) = \sigma(\bar{y})$. So we obtain the expression

$$\Delta_x^2 u(\bar{x}) \Delta_y^2 u(\bar{y}) = \sigma(\bar{x})\sigma(\bar{y}),$$

where Δ_x is the Laplace operator, operating in the variable $\bar{x}$.

Once the mathematical expectation from the both parts of the expression has been taken, we obtaine the equation for the unknown covariance function

$$\Delta_x^2 \Delta_y^2 v(\bar{x}, \bar{y}) = K(\bar{x}, \bar{y}).$$

Let us now rewrite it as a system of two equations

$$\Delta_x^2 w(\bar{x}, \bar{y}) - K(\bar{x}, \bar{y}), \quad \Delta_y^2 v(\bar{x}, \bar{y}) = w(\bar{x}, \bar{y}). \tag{4.26}$$

The boundary conditions for this system are obtained in the following way:

$$w(\bar{x}, \bar{y})\big|_{x \in \partial G} = \Delta_y^2 \mathrm{E}\{u(\bar{x})u(\bar{y})\}\big|_{x \in \partial G} = \varphi(\bar{x})\mathrm{E}\{\Delta_y^2 u(\bar{y})\} = \varphi(\bar{x})\mathrm{E}\{\sigma(\bar{y})\}$$

$$\Delta_x w(\bar{x}, \bar{y})\big|_{x \in \partial G} = \Delta_x \Delta_y^2 \mathrm{E}\{u(\bar{x})u(\bar{y})\}\big|_{x \in \partial G} = \psi(\bar{x})\mathrm{E}\{\Delta_y^2 u(\bar{y})\} = \psi(\bar{x})\mathrm{E}\{\sigma(\bar{y})\}$$

$$\begin{aligned} v(\bar{x}, \bar{y})\big|_{y \in \partial G} &= \mathrm{E}\{u(\bar{x})u(\bar{y})\}\big|_{y \in \partial G} = \varphi(\bar{y})\mathrm{E}\{u(\bar{x})\} \\ \Delta_y v(\bar{x}, \bar{y})\big|_{y \in \partial G} &= \Delta_y \mathrm{E}\{u(\bar{x})u(\bar{y})\}\big|_{y \in \partial G} = \psi(\bar{y})\mathrm{E}\{u(\bar{x})\}. \end{aligned} \tag{4.27}$$

To define the latter two conditions we are to solve the equations

$$\Delta^2 \mathrm{E}\{u(\bar{x})\} = \mathrm{E}\{\sigma(\bar{x})\}, \quad \mathrm{E}\{u\}\big|_{\partial G} = \varphi(\bar{x}), \quad \Delta \mathrm{E}\{u\}\big|_{\partial G} = \psi(\bar{x}).$$

We can solve each of the equations (4.26) using the 'walk on circles' algorithm of the Monte Carlo method (see Section 4.3).

Consider the second equation of the system (4.26). It is obvious that its solution satisfies the following integral relations

$$\begin{aligned} v(\bar{x}, \bar{y}_0) = {} & \frac{1}{2\pi r} \int\limits_{S(\bar{y}_0, r)} v(\bar{x}, \bar{s})d\bar{s} - \frac{r^2}{4} \int\limits_{S(\bar{y}_0, r)} \Delta_s v(\bar{x}, \bar{s})d\bar{s} \\ & + \frac{1}{2\pi} \int\limits_0^{2\pi} \int\limits_0^{r} z \frac{r^2 - z^2(\ln(r/z)+1)}{4} w(\bar{x}, \bar{y}_0 + (z\cos\varphi, z\sin\varphi))d\varphi dz, \end{aligned}$$

$$\Delta_y v(\bar{x},\bar{y}_0) = \frac{1}{2\pi r}\int\limits_{S(\bar{y}_0,r)} \Delta_s v(\bar{x},\bar{s})d\bar{s}$$
$$-\frac{1}{2\pi}\int\limits_0^{2\pi}\int\limits_0^{r} z\ln(r/z)w(\bar{x},\bar{y}_0+(z\cos\varphi, z\sin\varphi))d\varphi dz,$$

where $S(\bar{y}_0,r)$ is the maximal circle with the centre at the point $\bar{y}_0$ entirely lying in G. Here r is the radius of S.

We have obtained the system of the integral equations

$$V = KV + F.$$

The random estimate of the solution is

$$\bar{\xi} = \sum_{n=0}^{N-1} Q_n P(\bar{y}_n) + Q_N P(\bar{y}_N), \tag{4.28}$$

where $\{\bar{y}_n\}_{n=0}^N$ is the 'walk on circles' Markov chain which terminates at the random point $\bar{y}_N$ in the ε-neighbourhood of the boundary, $\{Q_n\}$ are the matrix weights:

$$Q_0 = \begin{bmatrix} 1 & 0 \\ 0 & 1 \end{bmatrix}, \quad Q_n = Q_{n-1}\begin{bmatrix} 1 & -r^2/4 \\ 0 & 1 \end{bmatrix},$$

where r is the radius of the maximum inscribed circle with the centre at the point $\bar{y}_{n-1}$, $P(\bar{y}_N) = P(\bar{y}_*)$, where $\bar{y}_*$ is the boundary point nearest to the point $\bar{y}_N$. The function P is determined by boundary conditions (4.27). This estimate is ε-biased.

The integrals $F_i(\bar{y}_n)$ are estimated by the Monte Carlo method using one random point. As a density in the segment $[0,r]$ we choose the function $f = 4\rho\ln(r/\rho)/r^2$, therefore the random estimate is

$$\begin{aligned} \tilde{F}_1(\bar{y}_n) &= \frac{r^2-\rho^2(\ln(r/\rho)+1)}{4}\cdot\frac{r^2}{4\ln(r/\rho)}w(\bar{x},\bar{y}_n+\rho\bar{\omega}), \\ \tilde{F}_2(\bar{y}_n) &= -\frac{r^2}{4}w(\bar{x},\bar{y}_n+\rho\bar{\omega}), \end{aligned} \tag{4.29}$$

where ρ is the random variable, distributed with the density f, $\bar{\omega}$ is the unit isotropic vector.

Thus, to find the solution of the problem $v(\bar{x}_0,\bar{y}_0)$, we need the values of function $w(\bar{x}_0,\bar{y}_n+\rho_n\bar{\omega}_n)$, $n=0,\dots,N-1$ and the value $\mathrm{E}\{u(x_0)\}$. They are found from the first equation of system (4.26) by means of the method under consideration, using only one realization of the Markov chain $\{\bar{x}_m\}_{m=0}^M$ for every w.

Here the 'walk on circles' is simulated starting from the point $\bar{x}_0$: $\{\bar{x}_m\}_{m=0}^M$. The random estimates for $w(\bar{x}_0,\bar{y}_n+\rho_n\bar{\omega}_n)$ by analogy with estimate (4.48) are

$$\bar{\zeta}_n = \sum_{m=0}^{M-1} Q_m\tilde{F}(\bar{x}_m) + Q_M P(\bar{x}_M). \tag{4.30}$$

Here we have the following random estimates for the components of the vector $\tilde{F}(\bar{x}_m)$:

$$\begin{aligned} F_1(\bar{x}_m, \bar{y}_n + \rho_n\bar{\omega}_n) &= \frac{r^2 - \rho^2(\ln(r/\rho)+1)}{4} \cdot \frac{r^2}{4\ln(r/\rho)} \\ &\quad \times K(\bar{x}_m + \rho\bar{\omega}, \bar{y}_n + \rho_n\bar{\omega}_n), \qquad (4.31) \\ F_2(\bar{x}_m, \bar{y}_n + \rho_n\bar{\omega}_n) &= -\frac{r^2}{4} K(\bar{x}_m + \rho\bar{\omega}, \bar{y}_n + \rho_n\bar{\omega}_n) \end{aligned}$$

are used, where ρ is the random variable distributed with the density $4\rho\ln(r/\rho)/r^2$, $\bar{\omega}$ is the unit isotropic vector. Furthermore, we see from (4.30), (4.31) that every $\bar{\zeta}_n$ can be calculated using only the Markov chain $\{\bar{x}_m\}_{m=0}^{M}$.

It is possible to realize this algorithm in the following way:

1. The Markov chain $\{\bar{y}_n\}_{n=0}^{N}$ and the random values ρ_n, distributed with the density f, are simulated. Thus, the points $\bar{y}_n + \rho_n\bar{\omega}_n$ are fixed.
2. The Markov chain $\{\bar{x}_m\}_{m=0}^{M}$ and values ρ_m, distributed with the density f, are simulated.
3. The integrals $F_i(\bar{x}_m)$ are estimated according to (4.31).
4. The random estimates $\bar{\zeta}_n$ are found using (4.30).
5. The integrals $F_i(\bar{y}_n)$ are estimated according to (4.29), where $w(\bar{x}_0, \bar{y}_n + \rho_n\bar{\omega}_n)$ defines the non-biased estimate $\bar{\zeta}_n$.
6. Solution of the initial problem is estimated from (4.28).

The efficiency of the given algorithm (A) was examined in comparison with the randomized algorithm B (see [47]) for numerical solving the problem of the oscillation of the plate

$$G = \{x_1, x_2 : \ 0 \le x_1, x_2 \le 1\}$$

simply supported at the boundary with the random field σ, having the spectral density

$$p(\lambda) = \frac{\alpha^2}{2\pi} e^{-\alpha|\lambda|}, \quad \lambda \in R^2.$$

Here, taking into consideration the piecewise linear boundaries, the boundary value problem is as follows:

$$\Delta^2 u(\bar{x}) = \sigma(\bar{x}), \quad u\big|_{\partial G} = \Delta u\big|_{\partial G} = 0.$$

System (4.26), (4.27) becomes

$$\begin{aligned} &\Delta_x^2 w(\bar{x},\bar{y}) = K(\bar{x},\bar{y}), \quad \Delta_y^2 v(\bar{x},\bar{y}) = w(\bar{x},\bar{y}), \\ &w(\bar{x},\bar{y})\big|_{x\in\partial G} = \Delta_x w(\bar{x},\bar{y})\big|_{x\in\partial G} = v(\bar{x},\bar{y})\big|_{y\in\partial G} = \Delta_y v(\bar{x},\bar{y})\big|_{y\in\partial G} = 0. \end{aligned}$$

The randomised spectral model (see [40])

$$\sigma(\bar{x}) = \sum_{k=1}^{M_\sigma} p_k^{1/2}[\xi_k \sin(\lambda_k\bar{x}) + \eta_k \cos(\lambda_k\bar{x})] \qquad (4.32)$$

was used in the algorithm B for simulating σ, where ξ_k, η_k are the independent standard Gaussian variables,

$$p_k = \int_{\Lambda_k} p(\lambda) d\lambda,$$

where $\{\Lambda_1, \dots, \Lambda_{M_\sigma}\}$ is some arbitrary partition of the plane R^2, and $\lambda_1, \dots, \lambda_{M_\sigma}$ are distributed in these parts in accordance with the densities $p_k(\lambda) = p(\lambda)/p_k$.

Let us prove that the variance of the algorithm A is less than the variance of the algorithm B. In the second case the random estimate of the solution is the product of two estimates

$$\bar{\xi}_x = \sum_{n=0}^{N-1} Q_n I_n \sigma(\bar{x}_n + \rho_n \bar{\omega}_n), \quad \bar{\xi}_y = \sum_{m=0}^{M-1} Q_m I_m \sigma(\bar{y}_m + \rho_m \bar{\omega}_m),$$

where

$$I_n \sigma(\bar{x}_n + \rho_n \bar{\omega}_n) = F(\bar{x}_n), \quad I_m \sigma(\bar{y}_m + \rho_m \bar{\omega}_m) = F(\bar{y}_m).$$

Their multiplication gives

$$\bar{\zeta} = \sum_{n=0}^{N-1} \sum_{m=0}^{M-1} Q_n Q_m I_n I_m \sigma(\bar{x}_n + \rho_n \bar{\omega}_n) \sigma(\bar{y}_m + \rho_m \bar{\omega}_m). \tag{4.33}$$

Let $\mathrm{M}[\bar{\zeta} \mid z]$ be the conditional expectation of the random variable ζ under the condition that the points $\bar{x}_n + \rho_n \bar{\omega}_n$ and $\bar{y}_m + \rho_m \bar{\omega}_m$ are fixed. Therefore, we can use a well-known relation

$$\mathrm{D}\bar{\zeta} = \mathrm{MD}[\bar{\zeta} \mid z] + \mathrm{DM}[\bar{\zeta} \mid z].$$

For the algorithm A estimate (4.33) is as follows:

$$\bar{\zeta}_* = \sum_{n=0}^{N-1} \sum_{m=0}^{M-1} Q_n Q_m I_n I_m K(\bar{x}_n + \rho_n \bar{\omega}_n, \bar{y}_m + \rho_m \bar{\omega}_m).$$

It is evident that the first term of sum expressing $\mathrm{D}\bar{\zeta}$ becomes zero and therefore,

$$\mathrm{D}\bar{\zeta}_* = \mathrm{DM}[\bar{\zeta}_* \mid z] = \mathrm{DM}[\bar{\zeta} \mid z] < \mathrm{D}\bar{\zeta}.$$

Thus, the variance of the algorithm A is less than the variance of the algorithm B.

Table 4.4 gives the results of comparison of the algorithms A and B. The covariance function v for the points $x = (1/2, 1/2)$ and $y = (1/2 + d, 1/2)$ was estimated. The calculations were performed for $\varepsilon = 0.01$, parameter $\alpha = 1$, a number of samples $N = 3000$. When σ is simulated according to (4.32), the simplest case of $M_\sigma = 1$ is used. Then $\Lambda = R^2$.

From this table it is seen that the computational cost of the algorithm A is $1.5 \div 3$ times less.

The method under consideration leads to the substantial efficiency of the calculations for those random field, whose simulation has the great computational cost. As an example, the comparison of the results of the calculations of the covariance function at the points $x = (0.5, 0.5)$ and $y = (0.5, 0.4)$ with the help of the algorithms A and B for parameters $\alpha = 0.1$, $\varepsilon = 0.01$ and $N = 10000$ is presented (see Table 4.5). Here ρ is a relative standard deviation error, s is the computational cost, M is a number of terms in formula (4.32).

Let us go to estimating the covariance function of the solution of the problem:

Table 4.4. $\alpha = 1$

d	A			B		
	$v \cdot 10^{-5}$	ρ, %	s	$v \cdot 10^{-5}$	ρ, %	s
0.0	1.37	3.5	0.18	1.26	4.3	0.26
0.1	1.32	4.8	0.39	1.25	6.0	0.58
0.2	1.04	3.6	0.22	1.16	6.2	0.63
0.3	0.87	5.5	0.51	0.82	7.2	0.83
0.4	0.39	5.7	0.52	0.38	11.2	1.95

Table 4.5. $\alpha = 0.1$

Algorithm	M_σ	$v \cdot 10^{-6}$	ρ, %	s
B	1	0.89	21	6.6
	2	1.013	17	5.4
	4	1.09	13.3	4.3
	6	1.22	12.22	4.5
	8	1.055	11.18	4.5
A		1.05	2.4	0.086

$$\Delta^2 u(\bar{x}) + \lambda(\bar{x})u(x) = g(\bar{x})$$

$$u\big|_{\partial G} = \varphi(\bar{x}), \quad \Delta u\big|_{\partial G} = \psi(\bar{x})$$

provided that $\lambda(\bar{x})$ and $g(\bar{x})$ are independent Gaussian fields. It is supposed, for the simplicity of the description, that

$$\mathrm{E}\lambda(\bar{x}) = \mathrm{E}g(\bar{x}) = 0, \quad \mathrm{D}\lambda(\bar{x}) = \mathrm{D}g(\bar{x}) = 1.$$

It is expedient to use here the 'walk on circles and circumferences' algorithm from Section 4.3. After averaging the product of the corresponding estimates for $u(\bar{x}_0)$ and $u(\bar{y}_0)$ with respect to the distributions of λ and g, we obtain the following ε-biased estimate for the covariance between $u(\bar{x}_0)$ and $u(\bar{y}_0)$:

$$\zeta = \sum_{i=1}^{n_N} \sum_{j=1}^{m_N} I_{n_j} I_{m_j} K_g(\bar{x}_{n_i} + \rho_{n_i}\omega_{n_i}, \bar{y}_{m_j} + \rho_{m_j}\omega_{m_j}) \mathrm{E}(Q_{n_i} Q_{m_j}),$$

where $\{\bar{x}_m\}_{m=0}^M, \{\bar{y}_n\}_{n=0}^N$ are the 'walk on circles and circumferences' beginning with the points $\bar{x}_0$ and $\bar{y}_0$ respectively, $K_g(\bar{x}, \bar{y})$ is the covariance function of the filed g, I_{n_i} and I_{m_j} are the auxiliary weights (see Section 4.3).

Further,

$$\begin{aligned} \mathrm{E}(Q_{n_i} Q_{m_j}) &= \mathrm{E}\Big\{\Big[\prod_{p=1}^{l_{n_i}} \frac{-\lambda(r_{k_p})}{c_1}\Big]\Big[\prod_{p=1}^{n_i - l_{n_i}} q_{t_p}\Big]\Big[\prod_{z=1}^{l_{m_j}} \frac{-\lambda(r_{k_z})}{c_1}\Big]\Big[\prod_{z=1}^{m_j - l_{m_j}} q_{t_z}\Big]\Big\} \\ &= J_{n_i m_j}\Big(-\frac{1}{c_1}\Big)^{l_{n_i} + l_{m_j}} \mathrm{E}\Big\{\prod_{s=1}^{l_{n_i} + l_{m_j}} \lambda(r_s)\Big\}. \end{aligned}$$

Therefore,

$$\zeta = \sum_{i=1}^{n_N} \sum_{j=1}^{m_M} J_{n_i m_j} \Big(-\frac{1}{c_1} \Big)^{l_{n_i}+l_{m_j}} \mathrm{E}\Big\{ \prod_{s=1}^{l_{n_i}+l_{m_j}} \lambda(r_s) \Big\}$$
$$\times K(\bar{x}_{n_i} + \rho_{n_i}\omega_{n_i}, \bar{y}_{m_j} + \rho_{m_j}\omega_{m_j}) \ .$$

The expression $\mathrm{E}\Big\{ \prod_{s=1}^{l_{n_i}+l_{m_j}} \lambda(r_s) \Big\}$ presents the moment of the field λ of the order $L = l_{n_i} + l_{m_j}$ which can be calculated by the formula [19]:

$$\mathrm{E}\Big\{ \prod_{s=1}^{L} \lambda(r_s) \Big\} = \sum_{\alpha_i \in \Lambda} \prod_{j=1}^{L/2} \mathrm{E}\{\lambda(r_{\alpha_{i,2j-1}})\lambda(r_{\alpha_{i,2j}})\},$$

where Λ is a set of $(n-1)!!$ permutations of the number $1, ..., n$, so that for every $\alpha_i \in \Lambda$ the following inequality is valid

$$\alpha_{i,2k-1} < \alpha_{i,2k}, \quad k = 1, ..., n/2 \ .$$

Here $\alpha_{i,p}$ is the p-th component of the permutation α_i. In Section 4.3 the condition $|\lambda| < c_1$ is assumed; here it changes to the following wider condition

$$\Big| \mathrm{E}\Big\{ \prod_{s=1}^{L} \lambda(r_s) \Big\} \Big| \Big/ c_1^L \le 1, \quad \forall L \ . \tag{4.34}$$

By analogy with Section 4.3 it is also possible to obtain the following conditions of the variance finitness especially for a biharmonic equation:

$$c_1 < 0.638c^*, \quad c^* > 2.4 \ .$$

Obviously, condition (4.34) is wider than the condition $|\lambda| < c_1$. Therefore the method constructed here is more effective than the dual randomization method related to the simulation of the fields λ and g (see Setion 1.6).

In the computations only moments of the order $L \le L_0$ are taken into account, i.e., the other moments change to zero. If $L_0 = 8$, then the following inequality holds:

$$\Big| \mathrm{E}\Big\{ \prod_{s=1}^{L} \lambda(r_s) \Big\} \Big| \le (\mathrm{D}\lambda)^{1/2}(L-1)!! \le (L-1)!! < (1.8)^L \ .$$

Hence, here inequality (4.34) is valid if $c_1 \ge 1.8$.. Note, that in the case $c_1 < 1.8$ condition (4.34) also can be valid because the points $\{r_s\}$ with the large probability are distributed in G at the essential distance, each one from each other and, therefore, the values

$$\mathrm{E}\{\lambda(r_{\alpha_{i,2j-1}})\lambda(r_{\alpha_{i,2j}})\}$$

can be essentially small.

Note that for the stochastic equation $\Delta^2 u + \lambda u = g$ the simpler problem of calculating the mean solution $\mathrm{E}u(x) = \tilde{u}(x)$ also is significant. The averaging of the initial problem gives the problem

$$\Delta^2 \tilde{u} + (\widetilde{\lambda u}) = \tilde{g}, \quad \tilde{u}\big|_{\partial G} = \varphi, \quad \Delta\tilde{u}\big|_{\partial G} = \psi \ .$$

It is interesting that the direct averaging of the probabilistic representation of the initial solution u shows that $\tilde{u}$ does not change after the transition from equation $\Delta^2 u + \lambda u = g$ to the equation $\Delta^2 u + \lambda u = \tilde{g}$ when λ and g are the independent random fields. The above considered Monte Carlo technique is used for calculating $\tilde{u}$ in the obvious way.

4.5. MONTE CARLO SOLUTION OF THE DIRICHLET PROBLEM FOR ELLIPTIC SYSTEMS WITH VARIABLE PARAMETERS

Let us consider the Dirichlet problem for the system of the equations:

$$\Delta u_i(r) + \sum_{j=1} c_{ij}(r)u_j(r) = -g_i(r), \quad u_i(r)\big|_\Gamma = \psi_i(r), \quad i = 1,...,m \tag{4.35}$$

in a domain $D \subset R^n$ with the boundary Γ, $r = (x_1, ..., x_n) \in D$. The conditions for the functions$\{c_{ij}, g_i, \psi_i\}$ and the boundary Γ to be regular are assumed to be fulfilled. These conditions ensure existence and uniqueness of the solution of problem (4.35) (see below), as well as the probabilistic and integral representations, which will be used below. The functions $\{c_{ij}, g_i, \psi_i\}$ may be complex-valued.

Let us transform (4.35) to the form:

$$\Delta u_i(r) = -\sum_{j=1}^{m} c_{ij}(r)u_j(r) - g_i(r), \quad u_i(r)\big|_\Gamma = \psi_i(r), \quad i = 1,...,m\ . \tag{4.36}$$

On the basis of the integral representation of the Dirichlet problem solution in the centre of a ball, it is possible to write for functions $\{u_i\}$ the special system of integral equations, which in $D\backslash\Gamma_\varepsilon$ has the form:

$$\begin{aligned} u_i(r) = {} & [1 - c_0 d^2(r)/2n] \int_{S(r)} \frac{[1 - c_0 d^2(r)/2n]^{-1}}{\omega_n d^{n-1}(r)} u_i(r(s))ds \\ & + \sum_{j=1}^{m} \frac{1}{m} \cdot \frac{c_0 d^2(r)}{2n} \int_{D(r)} \frac{m \cdot c_{ij}(r')}{c_0} 2nd^{-2}(r)G_r(r')u_j(r')dr' \\ & + \int_{D(r)} G_r(r')g_i(r')dr', \quad i = 1,...,m, \quad r \in D\backslash\Gamma_\varepsilon, \end{aligned} \tag{4.37}$$

while in Γ_ε it transforms into the identity:

$$u_i(r) \equiv u_i(r), \quad r \in \Gamma_\varepsilon\ .$$

Here $d(r)$ is the distance from the point r to the boundary Γ; $S(r)$, $D(r)$ are the sphere and the ball of radius $d(r)$ centred at the point r; $G_r(r')$ is the ball Green function for the Laplace equation in the centre of the ball $D(r)$:

$$G_r(r') = \frac{1}{(n-2)\omega_n}\left(\frac{1}{|r - r'|^{n-2}} - \frac{1}{d^{n-2}(r)}\right), \quad |r - r'| \le d(r)\ .$$

The value c_0 has to satisfy the inequality

$$c_0 d^2_{\max}/(2n) < 1\ . \tag{4.38}$$

If one more integral-valued coordinate i is introduced in the phase spase, then it is possible to consider system (4.37) as an integral equation with respect to the unknown function $u(x) \equiv u(r,i) \equiv u_i(r)$:

$$u = Ku + r\ . \tag{4.39}$$

Here the sum with respect to i is considered as the integral with respect to the corresponding discrete measure.

The standard Monte Carlo estimate for solving an integral equation, in accordance with the Neumann expansion, has the form (see Section 1.2)

$$\xi_0 = \xi_{x_0} = \xi_{r_0,i_0} = \sum_{k=0}^{N} Q_k h_{i_k}(r_k), \quad \mathrm{E}\xi_0 \equiv u(x_0) = u_{i_0}(r_0),$$

where $\{r_k\}$ is a terminating Markov chain and $\{Q_k\}$ are the corresponding weights. In accordance with the form of integral equation (4.39) it is possible to construct here the standard estimate on the 'walk on spheres and balls'. Here, with the probability $1 - c_0 d^2(r_k)/(2n)$ a new point r_{k+1} is chosen uniformly over the sphere $S(r_k)$, while the weight Q_k is multiplied by the quantity

$$q(r_k, c_0) = [1 - c_0 d^2(r_k)/(2n)]^{-1}$$

and a number i_{k+1} is preserved; with the probability $c_0 d^2(r_k)/(2n)$ a point r_{k+1} is chosen within the ball $D(r_k)$ with respect to the density $2nd^{-2}(r_k)G_{r_k}(r')$ and number i_{k+1} is simulated uniformly in $\overline{1,m}$, while the weight Q_k is multiplied by the quantity $c_{i_k,i_{k+1}}(r_{k+1}) \times m/c_0$. As a result

$$Q_k = \left[\prod_{i=1}^{m_k} c_{l_{k_i-1},l_{k_i}}(r_{k_i}) \cdot m/c_0\right]\left[\prod_{i=1}^{k-m_k} q(r_{t_i-1}, c_0)\right] .$$

Here $\{r_{t_i}\}$, $i = 1, ..., k - m_k$, are the points, chosen on the spheres, while $\{r_{k_i}\}$, $i = 1, ..., m_k$, are the points, chosen inside the balls, and $\{l_k\}$ are the chosen numbers. The functions $h_i(r)$ are calculated exactly or in a randomized manner (see Section 3.2), when the chain finds itself inside Γ_ε it is terminated, and the solution estimate multiplied by the weight is added to the sum in the counter. The actual estimate ξ_ε is obtained by the use $h_i(r) = \psi_i(P(r))$ for $r \in \Gamma_\varepsilon$, where $P(r)$ is the boundary point closest to r.

Lemma 4.4. *Provided*

$$c_0 \geq m \cdot |c_{ij}(r)|, \quad i,j = 1, ..., m, \ \forall r \in D; \quad c_0 < 2nc^*/\alpha_n^2,$$

where $-c^*$ is the first eigenvalue of the Laplacian operator for the domain D, and α_n is the first positive root of the Bessel function $J_{(n-2)/2}(x)$, then the Neumann series for equation (4.39) converges.

Proof is similar to the proof of lemma 3.1. □

Note, that the condition $c_0 < 2nc^*/\alpha_n^2$ implies inequality (4.38).

Theorem 4.3. *Under the hypotheses of Lemma 4.4 there exists an unique bounded solution to equation (4.39) that can be represented by the corresponding Neumann series and coincides with the solution of problem (4.35).*

Proof is similar to the proof of Theorem 3.1. □

Theorem 4.4. *If the first derivatives of the solution of problem (4.39) are bounded in D, then, provided the conditions of Lemma 4.4 are satisfied,* $\mathrm{E}\xi_\varepsilon = u_\varepsilon(r,i)$ *exists and*

$$|u(r,i) - u_\varepsilon(r,i)| \leq \text{const} \cdot \varepsilon, \quad r \in D, \quad \varepsilon > 0, \quad i = 1, \ldots, m \, .$$

Proof is similar to the proof of Theorem 3.2. □

Now let us consider the question about uniform with respect to ε boundness of the variance $\mathrm{D}\xi_\varepsilon$.

Theorem 4.5. *Provided* $g_i \equiv 0$, *and*

$$c_0 \geq m \cdot |c_{ij}(r)|, \quad \forall r \in D, \quad i,j = 1, \ldots, m;$$

$$\frac{c_0}{c} < t(n) = \min_{x \in [0, \alpha_n]} \left\{ \frac{2n}{x^2} \left(1 - \frac{\Gamma(n/2) J_{(n-2)/2}(x)}{(x/2)^{(n-2)/2}} \right)^{1/2} \right\},$$

we have

$$\mathrm{D}\xi_\varepsilon < c_d < +\infty, \quad \forall \varepsilon > 0 \, .$$

Proof is similar to the proof of Theorem 3.3. □

Computations gave the following values: $t(2) = 0.5$, $t(3) = 0.488...$, $t(4) = 0.465...$, $t(10) = 0.336...$.

In the case of nonhomogeneous system (4.35), with the aim to extend the condition of the finitness of the variance, let us modify the estimate as follows. We shall calculate the function $h(r,i)$ only for the step $(r,i) \to (r',i)$ onto the sphere, however, with the weight $q(r, c_0)$, i.e., $h(r,i) \to h_1(r,i)$:

$$h_1(r_{k_i-1}, l_{k_i-1}) = 0, \quad h_1(r_{t_i-1}, l_{t_i-1}) = q(r_{t_i-1}, c_0) h(r_{t_i-1}, l_{t_i-1}).$$

Thus, instead of the estimate ξ_ε we consider the estimate

$$\xi_{\varepsilon,1} = \sum_{k=0}^{N} Q_k h_1(x_k) = \sum_{i=1}^{N-m_N} Q_{t_i} h(r_{t_i-1}, l_{t_i-1}) + Q_N \psi(P(r_N), l_N) \, .$$

The repeated averaging yields: $\mathrm{E}\xi_{\varepsilon,1} = \mathrm{E}\xi_\varepsilon$.

Theorem 4.6. *Under the hypotheses of Theorem 4.5 we have*

$$\mathrm{D}\xi_{\varepsilon,1} < C_{d,1} < +\infty, \; \forall \varepsilon > 0, \; r \in D.$$

Proof is similar to the proof of Theorem 3.4. □

To generalize the constructed algorithm, it is possible to simulate the integral-value coordinate with respect to the probabilities $p_j(r_{k+1})$, proportional to $|c_{ij}(r_{k+1})|$; here Lemma 4.4 and Theorems 4.3-6 are valid after the change of the condition $c_0 \geq m|c_{ij}(r)|$ for the condition

$$c_0 \geq \sum_j |c_{ij}(r)|, \quad i = 1, ..., m; \quad r \in D .$$

With the use of Theorem 3.1a it is possible to generalize this condition for Theorems 4.3,4 in the following manner:

$$c^* > \sum_j |c_{ij}(r)|, \quad i = 1, ..., m; \quad r \in D \tag{4.40}$$

On the basis of Theorem 4.3 and the Fredholm alternative one can conclude that condition (4.40) quadrantees existence and uniqueness of the solution of problem (4.35).

In addition, it is possible to construct the vector algorithm for solving problem (4.35), on the basis of system (4.37) without randomization with respect to j; here values m are cancelled. The vector estimate of the solution $u(r)$ has the form:

$$\zeta_0 = \sum_{k=0}^{N} Q_k H(r_k), \quad u(r_0) = \mathrm{E}\zeta_0,$$

where $\{r_k\}$ is the chain of 'walk on spheres and balls', Q_0 is the unitary matrix, and the weight Q_{k+1} is determined as follows: if r_{k+1} is chosen onto the sphere, then the matrix Q_k is multiplied by the scalar $q(r_k, c_0)$, while if r_{k+1} is chosen within the ball, then Q_k is multiplied from the right by the matrix $C(r_{k+1})/c_0$, where $C(r) = \{c_{ij}(r)\}$. As a result

$$Q_k = \left[\prod_{i=1}^{m_k} C(r_{k_i})/c_0\right]\left[\prod_{i=1}^{k-m_k} q(r_{t_i-1}, c_0)\right] .$$

The ε-biased estimate has the form:

$$\zeta_{\varepsilon,0} = \sum_{k=0}^{N-1} Q_k H(r_k) + Q_N \psi(P(r_N)),$$

where $\psi(P) = (\psi_1(P), ..., \psi_m(P))$. For the vector estimate Lemma 4.4 and Theorems 4.3-6 are valid after the change of the condition $c_0 \geq m|c_{ij}(r)|$ for the condition $c_0 \geq \sum |c_{ij}(r)|$, $i = 1, ..., m$, $r \in D$. Theorems 4.3,4 can be further generalized as mentioned above.

It is easy to note that it is possible to obtain the vector estimate from the original scalar estimate by the way of the conditional averaging with respect to j, reducing the variance. The vector estimate can be essentially effective if the dependent estimating u-components is necessary.

It is possible to extend the constructed algorithms to the case of the Dirichlet problem for the integro-differential equation:

$$\Delta u(r, \lambda) = -\int_\Lambda u(r, \lambda')c(r, \lambda, \lambda')d\lambda' - g(r, \lambda), \quad u(r, \lambda)\big|_\Gamma = \psi(r, \lambda) .$$

Here, if in the process of 'walk on spheres and balls' a new point r_{k+1} is chosen within the ball, then a new value λ_{k+1} is chosen with respect to the density $p(r_{k+1}, \lambda_k, \lambda')$, while the weight is multiplied by the quantity $c(r_{k+1}, \lambda_k, \lambda_{k+1})/(c_0 p(r_{k+1}, \lambda_k, \lambda_{k+1}))$. Lemma 4.4 and Theorems 4.3-6 are here valid after the change of the condition $c_0 \geq m|c_{ij}(r)|$ for the condition

$$c_0 \geq |c(r,\lambda,\lambda')|/p(r,\lambda,\lambda'),\ \forall r \in D,\ \lambda,\lambda' \in \Lambda$$

and can be further generalized as mentioned above. For instance, it is possible to use in Lemma 4.4 the condition

$$c^* > c(r,\lambda,\lambda')|/p(r,\lambda,\lambda'),\ \forall r \in D,\ \lambda,\lambda' \in \Lambda\ .$$

We can here consider

$$p(r,\lambda,\lambda') = \frac{|c(r,\lambda,\lambda')|}{\int_\Lambda |c(r,\lambda,\lambda')| d\lambda'}\ .$$

Therefore, the condition

$$\int_\Lambda |c(r,\lambda,\lambda')| d\lambda' < c^*,\ \forall r \in D,\ \lambda \in \Lambda$$

guarantees existence and uniqueness of the solution of the Dirichlet problem for the integro-differential equation.

Now let us consider in the domain $D \in R_3$ with the boundary Γ the first boundary-value problem for the system of the equations:

$$\Delta u_1 + \gamma(r)\frac{\partial u_2}{\partial x} = -g, \quad \Delta u_2 = 0. \tag{4.41}$$

Here

$$u_1\big|_\Gamma = \psi_1, \quad u_2\big|_\Gamma = \psi_2 \tag{4.42}$$

and the standard restrictions for Γ, γ and g are supposed to be fulfilled.

The first equation (4.41) is the Poisson equation with respect to the function u_1. The corresponding randomized ε-biased Monte Carlo estimate on the 'walk on spheres and balls' $\{r_i\}$ has the form:

$$\eta_\varepsilon = \sum_{i=0}^{N} \frac{1}{6} d^2(r_i)\left[\gamma(\rho_i)\frac{\partial u_2(\rho_i)}{\partial x} + g(\rho_i)\right] + \psi_1(P(r_N)), \tag{4.43}$$

where ρ_i is a random point, chosen in the ball $D(r_i)$ with respect to the distribution, proportional to $G_{r_i}(\rho)$ (see section 3.3).

Further, it is known that the ε-biased estimate of the derivative $\partial u_2(\rho)/\partial x$ can be constructed on the 'walk on spheres'

$$\{\rho^{(n)}\}; \quad n = 0,1,\ldots,N_\rho; \quad \rho^{(0)} = \rho,$$

by the way of differentiating the noncentral spherical weight for the sphere $S(\rho)$. This estimate is as follows (see Section 3.3):

$$\xi'_\varepsilon(\rho) = \frac{3a}{d(\rho)}\psi_2(\rho^{(N_\rho)}),$$

where $a = \cos\alpha$, and α is the angle between the radius-vector ρ and the x-axis. Thus, it is possible to use instead of η_ε the additionally randomized ε-biased estimate:

$$\tilde{\eta}_\varepsilon = \sum_{i=0}^{N} \frac{1}{6} d^2(r_j)[\gamma(\rho_i)\xi'_\varepsilon(\rho_i) + g(\rho_i)] + \psi_1(P(r_n)) \,. \tag{4.44}$$

Here, random quantities $\{\xi'_\varepsilon(\rho_i)\}$ are conditionally independent for fixed $\{r_i\}$, because they are constructed on the different trajectories of the 'walk on spheres'. Owing to the multiplier $1/d(\rho_i)$ the quantity $\mathrm{D}\xi'_\varepsilon(\rho_i)$ for fixed r_i diverges logarithmically [39] and it acquires the order of the quantity $|\ln\varepsilon|$, if ξ'_ε is supposed to be zero for $\rho_i \in \Gamma_\varepsilon$. After this modification the estimate $\tilde{\eta}_\varepsilon$ remains ε-biased and its variance becomes the finite quantity of the order $|\ln\varepsilon|$. The cost of the constructed estimate is the quantity of the order $|\ln\varepsilon|^3$, because here it is necessary to simulate the additional trajectories. However, it is possible to construct the algorithm with the cost of the order $|\ln\varepsilon|^2$, using the 'absorption estimate' [40]. For this purpose, at each point $r_i \in D\backslash\Gamma_\varepsilon$ the termination of trajectory is simulated with respect to some probability p. If the trajectory does not terminate, then the weight is multiplied by the quantity $1/(1-p)$, while if the trajectory terminates, then the weight is multiplied by $1/p$ and the quantity, summarized in (4.44), is multiplied by the weight and added to the result. In accordance with Section 3.4 the value

$$p = c_0 d^2(r_i)/6,$$

with $c_0 < 0.488c^*$, is expedient here.

On the basis of Section 3.4 it is possible to extend the constructed algorithm to the case, when the second equation from (4.41) has the form:

$$\Delta u_2 + c(r)u_2 = 0,$$

because in Section 3.4 the ball Green function of the Laplace operator is used. The same circumstance allows one to estimate simultaneously derivatives with respect to the coordinates, as in Section 3.3.

Chapter 5

Monte Carlo methods with calculating parametric derivatives in the radiation transport theory

5.1. MONOVELOCITY TRANSFER PROCESS

The transfer process is regarded here as the stationary Markov chain whose states are particle collisions with atoms of the medium. When a collision takes place, absorption or scattering may occur; i.e. the particle may travel no further, or it may leave the point of collision in a new direction. The straight path between two successive collisions is is called a free path. The angle between the previous and the new direction is called the scattering angle. If the particle undergoes absorption or escapes from the medium, its trajectory is terminated. For the sake of simplicity, stationary problem will be considered without taking into account change of energy. We introduce the notation:

$r = (x, y, z)$: position vector ($r = |r|$);

$\omega = (a, b, c)$: direction of propagation of the radiation ($a_2 + b^2 + c^2 = 1$);

$\mu = \omega \cdot \omega'$: cosine of the scattering angle;

$g(\mu, r)$: phase function (indicatrix), normalized by

$$\int_{-1}^{+1} g(\mu, r) d\mu = 1;$$

$\sigma_s(r)$: scattering coefficient (cross section);

$\sigma_c(r)$: absorption coefficient;

$\sigma = \sigma_s + \sigma_c$: extinction coefficient;

$f(r, \omega)$: collision density;

$\Phi(r, \omega)$: particle flux (irradiance);

$\varphi(x) = f^*(x)$: collision 'importance';

$\Phi^*(x)$: particle 'importance'.

By definition, $\Phi(r, \omega) dS \cdot d\omega$ is the average number of particle that cross a plane element dS perpendicular to the direction ω, in a direction $\omega' \in [\omega, \omega + d\omega]$. It is well known that

$$f(r, \omega) = \sigma(r) \cdot \Phi(r, \omega). \tag{5.1}$$

The particle flux across an arbitrary plane element dS is defined by

$$\Phi_s(r,\omega) = |n_s\omega| \cdot \Phi(r,\omega), \tag{5.2}$$

where n_s is the normal to dS. Hence, the average number of particles that cross a sufrace S in the direction $\omega \in \Omega_k$ is given by

$$\int_S dS \int_{\Omega_k} \Phi[r(S),\omega]|\omega \cdot n_s|d\omega, \tag{5.3}$$

where Ω_k is an arbitrary subset of the sphere of directions.

We now define probability distributions of elements of the transfer process.

The free-path length l in the direction ω has the density function

$$f_l(t) = \sigma[r(t)] \exp\left\{ - \int_0^1 \sigma[r(t_1)]dt_1 \right\}, \tag{5.4}$$

where $r(t) = r_0 + t\omega$, r_0 is the initial position. The quantity $\tau(t) = \int_0^1 \sigma[r(t_1)]dt_1$ is called the optical depth of $[r_0, r(t)]$. To normalize Eq. (5.4), we suppose that the medium is bounded by a convex surface outside which $\sigma = \sigma_c \neq 0$. The cosine of the scattering angle at the point r has the distribution density $g(\mu, r)$. For several types of scattering with cross sections $\sigma_s^{(i)}(r)$ and indicatrixes $g_i(\mu, r)$, we put

$$\sigma_s(r) = \sum_i \sigma_s^{(i)}(r), \quad g(\mu, r) = \frac{\sum_i \sigma_s^{(i)}(r) g_i(\mu, r)}{\sigma_s(r)},$$

The probabilities of absorption and scattering at a point r are defined by

$$p(r) = \sigma_c(r)/\sigma(r), \quad g(r) = \sigma_s(r)/\sigma(r).$$

It follows from Eq. (5.4) that the probability of particles escape from the medium is given by

$$P[l > t^*(r_0,\omega)] = \exp\{-\tau[t^*(r_0,\omega)]\},$$

where $t^*(r_0,\omega)$ is the distance from r_0 to the surface in the direction ω.

The Monte Carlo method for solving problems of transfer theory consists of computational simulation of particle trajectories according to the following scheme:

1) The original position is simulated by to a source distribution $\psi(r,\omega)$.
2) The free-path length l is simulated.
3) The escape from the medium is examined.
4) The coordinates of the new collision are calculated:

$$x = x' + al, \quad y = y' + bl, \quad z = z' + cl$$

5) The type of collision (absorption or scattering) is simulated.
6) The cosine of the scattering angle μ is simulated.
7) The coordinates of a new direction are calculated:

$$a = a'\mu - (b' \sin\varphi + a'c' \cos\varphi)[(1-\mu^2)/(1-c'^2)]^{1/2}$$

$$b = b'\mu + (a'\sin\varphi - b'c'\cos\varphi)[(1-\mu^2)/(1-c'^2)]^{1/2}$$
$$c = c'\mu + (1-c'^2)\cos\varphi[(1-\mu^2)/(1-c'^2)]^{1/2},$$

in which φ is the azimuth of the scattering; φ is isotropic, so we have $\varphi = 2\pi\alpha$. Here and subsequently, α denotes a random number uniformly distributed between 0 and 1. However, more effective is a rejection method in which $\cos\varphi$ and $\sin\varphi$ are simulated as the coordinates of the isotropic unit vector,

i) $W_1 = l - 2\alpha_1, \quad W_2 = 1 - 2\alpha_2$
ii)$d = W_1^2 + W_2^2;$ if $d > 1$, go to i);
iii) otherwise

$$\cos\varphi = W_1 \cdot d^{-1/2}, \sin\varphi = W_2 \cdot d^{-1/2}$$

8) Go to 2).

If escape occurs in 3) or absorption in 5), then isstored and a new trajectory is simulated. The most complicated step is the simulation of the free-path length if σ is not constant for the medium. By integration of Eq. (5.4) we obtain the distribution function of l,

$$F_l(t) = 1 - \exp[-\tau(t)], \quad t > 0.$$

Thus l could be obtained from

$$\tau(l) = \int_0^l \sigma[r(t)]dt = -\ln\alpha,$$

which is easy to solve if $\sigma[r(t)]$ is a step function. Therefore the medium is usually divided into sufficiently small regions each having constant σ. Some algorithms for calculating distances to the boundary of these regions and simulations of l are given in [28]. Another method for an arbitrary $\sigma(r)$ – 'the maximal cross section method' – is considered below.

Let us consider the question of how numerical characteristics of the transfer process could be estimated by computational simulation. The integral

$$\int_{D_i} f(x)dx, \quad x = (r, \omega)$$

is estimates by the average number of particle collisions in the domain D_i.

If we associate with a collision a weight $[\sigma(r)]^{-1}$, then from Eq. (5.1) we obtain an estimate of the integral

$$\int_{D_i} \Phi(x)dx.$$

Another well-known estimate of the integral

$$\int_{R_i} dr \int \Phi(r,\omega)d\omega,$$

where R_i is the coordinate space, and Ω is the space of directions ω, is the 'free-path estimate' [53]. It is based on the fact that this integral can be shown to be equal to the average length L_i of the particle trajectory inside the region R_i. Indeed, intuitively, the free-path length is equal to σ^{-1}; hence, the average number of collisions inside the region R_i is equal to $\sigma \cdot L_i$. Consequently, the average number of collisions with weight σ^{-1} is L_i.

This algorithm is particularly convenient for regions of small volume but of great length, i.e., when the region is crossed frequently but collisions seldom occur. In [40], it is shown that in the case of the weight simulation technique, an unbiased free-path estimate can be constructed using the weights $\{Q_n\}$. This means that the part of the particle run lying in D_i is taken into account with a weight corresponding to the end of this run.

The integral of Eq. (5.3) is estimated by the average number of particles that cross a surface S in the direction $\omega_i \in \Omega_i$. If we assign to these crossings the weigh $(n_s \cdot \omega)^{-1}$, we obtain by Eq. (5.2) an estimate of the integral

$$\int_s d\Phi \int_{\Omega_i} \Phi[r(s), \omega] d\omega.$$

Thus the mean values of $f(x)$ and $\Phi(x)$ can be estimated. For large regions, this estimate is not sufficiently accurate; but for small regions, the statistical error is great. Therefore, the local radiance estimates, that can be introduced on the basis of the integral equation of transfer, are of great importance [24,28].

It should be noted that to solve some physical problems, it is sufficient to obtain the integral characteristics of the transfer process.

Now let us consider the maximal cross section method for simulating the free-path length. Let $W(\omega, \omega', r)$ be the distribution density of direction ω after scattering at the point r when ω' is the direction before the scattering.

The particle flux satisfies

$$(\omega, \operatorname{grad} \Phi) + \sigma(r)\Phi(r, \omega) = \int \Phi(r, \omega')\sigma_s(r)W(\omega, \omega', r)d\omega' + \Phi_0(r, \omega). \tag{5.5}$$

After adding Eq. (5.5) and the equality

$$[\sigma_m - \sigma(r)]\Phi(r, \omega) = \int \Phi(r, \omega')[\sigma_m - \sigma(r)]\delta(\omega' - \omega)d\omega'.$$

and combining the right-hand integrals, we may consider the equations obtained as the equation of transfer for a fictitious medium in which σ_m is the extinction coefficient, $\sigma_s(r)$ is the cross section of scattering with indicatrix $W(\omega, \omega', r)$, and $[\sigma_m - \sigma(r)]$ is the cross section of the scattering with constant ω'. Thus the direct simulation of this process obviously yields the desired simulation of the free-path length l.

This justification shows how this method could be applied to certain parts of the system. It also becomes clear how such a simulation can be combined with the weight method.

It is well known that the average number of collisions is given by $(\sigma, \Phi) = \int\int \sigma(r)\Phi(r, \omega)dr d\omega$. Consequently, the average number of collisions for the transformed equation is equal to (σ_m, Φ).

Here we obtain the integral equation of the second kind for the particle collision density.

It was mentioned above that the transfer process is regarded as a Markov chain of particle collisions in a medium. We obtain now a the transition density $k(x', x)$ of this chain, proceeding from the fact that the quantity $k(x', x)dx$ is equal to the number of particles leaving a collision at x' and having the next collisions in $(x, x + dx)$.

Let us consider a collision *at* the phase point $x' = (r', \omega)$ and the right circular cylinder of height dh and base perpendicular to ω, with area dS about the point r'. The probability that the particle will cross the area element dS after scattering is equal to

$$\frac{\sigma_s(r')}{\sigma(r')}\frac{g(\mu)}{2\pi}\frac{|r-r'|^2}{dS}, \quad \text{where} \quad \mu = \frac{\omega' \cdot (r-r')}{|r-r'|}. \tag{5.6}$$

Indeed, the element of solid angle is given by

$$d\omega = d\mu d\varphi.$$

We have

$$P(\mu_0 \leq \mu\mu_0 + d\mu;\ \varphi_0 \leq \varphi \leq \varphi_0 + d\varphi) = g(\mu_0)d\mu d\varphi/2\pi,$$

because μ and φ are independent.

Hence $g(\mu)/2\pi$ is the density-of-scattering distribution for the solid angle. Then, keeping in mind that $dS/|r-r'|^2$ is the value of the solid angle, corresponding to dS, we obtain Eq. (5.6).

Further, we have by Eq. (5.4) that the conditional probability (under the condition $\omega' \in d\omega$, where ω' is a new direction) of the collision in the cylinder is equal to $\sigma(r)\exp\{-\tau(r,r')\}$, where $\tau(r,r')$ is the optical length between r' and r.

Hence the absolute probability of collision in this cylinder is given by

$$\frac{\sigma_s(r')g(\mu)\exp[-\tau(r',r)]\sigma(r)}{\sigma(r')2\pi|r-r'|^2}dh\,dS. \tag{5.7}$$

We obtain the absolute density of distribution of r by dividing Eq. (5.7) by $dh\,dS$, i.e., volume of the cylinder. To obtain $k(x',x)$, we note that if r is fixed then ω is determined uniquely.

Therefore,

$$k(x',x) = \frac{\sigma_s r' g(\mu)\exp[-\tau(r',r)]\sigma(r)}{\sigma(r')\cdot 2\pi|r-r'|^2}\delta\left(\omega - \frac{r-r'}{|r-r'|}\right) \tag{5.8}$$

where $\mu = \omega' \cdot (r-r')/|r-r'|$. The expressions just obtained can be verified in the following way. The expressions Eq. (5.7) in polar coordinates with the center at r' yields the densities for μ, φ and l. Let $f_n(x)$ be a collision density of the n-th order. Clearly, $f_n = Kf_{n-1}$, where K is the integral operator with the kernel function $k(x',x)$, i.e.,

$$f_n(x) = \int_X k(x',x)f_{n-1}(x)dx'.$$

Since $f = f_0 + f_1 + f_2 = \ldots$, we have

$$f = \sum_{n=0}^{\infty} K^n\psi, \quad \text{where} \quad \psi = f_0. \tag{5.9}$$

Consequently,

$$f(x) = \int_X k(x',x)f(x')dx' + \psi(x), \quad f = Kf + \psi, \tag{5.10}$$

which is called an integral equation of transfer.

It is well known that series Eq. (5.9) converges if, for some n_0, $\| K^{n_0} \| < 1$. It is easy to show that $K \in (L_1 \to L_1)$.

Therefore, by Eq. (5.8)

$$\int_X k(x', x)dx \leq \sigma_s(r')/\sigma(r') = q(r').$$

Hence $\| K \|_{L_1} \leq \sup_{r'} q(r')$.

Thus, if $q(r') \leq q_0 < 1$, then $\| K \|_{L_1} < 1$.

In practice, it may appear that $\| K \|_{L_1} = 1$, but $\| K^{no} \|_{L_1} < 1$ for sufficiently large n_0.

It is easy to see that for a bounded medium $\| K^2 \|_{L_1} < 1$ even if $q(r') = 1$. The second power appears here because of the assumption that the medium is bounded by a convex surface outside which $\sigma(r) = \sigma_c(r) > 0$.

It is reasonable to consider the integral equation of transfer in the space N_1 of generalized measure densities of bounded variation, containing the space L_1. This is expedient when the distribution of the initial collision has a generalized density $\psi(x)$. For instance, for the monodirectional source, the density has the factor $\delta(\omega - \omega_0)$. It is esy to see that $K \in [N_1 \to N_1]$ and $\| K \|_{N_1} < 1$ if $q_0 < 1$.

The Monte Carlo method is usually used to estimate linear functionals of the form

$$I_h = (f, h) = \int_X f(x)h(x)dx.$$

The integral exists if $f \in L_1$, $h \in L_\infty$ or $f \in N_1, h \in C(X)$. For example, to estimate the integral $\int_D f(x)dx$, where D is a bounded domain, we must put

$$h(x) = \begin{cases} 1 & x \in D, \\ 0 & x \notin D. \end{cases}$$

Because of the elements of the transfer equation are nonnegative, the Monte Carlo method is applicable for an arbitrary $h \geq 0$ provided that $(f, h) < +\infty$. If $\{x_n\}$ is the chain of collisions, then

$$I_h = \mathrm{E}\xi, \quad \xi = \sum_{n=0}^{N} h(x_n).$$

To solve the problem under study, it is possible to simulate another Markov chain if only $r(x', x)$ contains a δ function as in Eq. (5.8). When the weight Q_n is constructed, these δ functions can be, formally, cancelled. It is easy to show, as in Section 1.2, that the estimate

$$\xi = \sum_{n=0}^{N} Q_n h(x_n)$$

is unbiased. Algorithms with $Q_n \neq 1$ are called weight methods. The weight method enables us to solve problems of the theory of transfer in various media by simulation of a particular Markov chain. By use of results of Chapter 1, in some case a weight method can be constructed that has less variance than the direct simulation method.

For example, we may multiply the weight of a particle by the scattering probability $\sigma_s(r)/\sigma(r)$ instead of simulating absorption or we may multiply the weight by $1 - \exp[-\tau(t^*)]$ instead of simulating escape. Another effective method of taking into account absorption implies $p(x'.x) = k(x'.x)$, corresponding to the medium with $\sigma = \sigma_s$, and $Q_n = \exp(-\sigma_c L_n)$, where L_n is the total run to x_n. In the case of 'δ-scattering' this method is 'importance sampling' and yields zero variance. It is easy to see from Eq. (1.29) that the variance is reduced but the time t required for computational simulation of the trajectory is increased. Consequently, if the quantity $t \cdot D\xi$ decreases, the modification is expedient (see Section 1.1).

The method of 'conjugated walks' is consider below. Let $I_p = (\Phi, p)$ be the function to be estimated, where p is some nonnegative function. It is well known that

$$I_p = (\Phi, p) = (\Phi^*, \Phi_0),$$

where Φ^* is the solution of the adjoint transfer equation

$$-(\omega, \operatorname{grad} \Phi^*) + \sigma\Phi^* = \int W(\omega', \omega, r)\sigma_s(r) \cdot \Phi^*(r, \omega')d\omega' + p,$$

with the boundary condition

$$\Phi^*(r, \omega) = 0 \quad \text{if} \quad r \in \Gamma, \quad \omega \cdot n < 0.$$

Hence for $\Phi^*(r, \omega) = \Phi_1^*(r, -\omega)$ we get

$$(\omega, \operatorname{grad} \Phi_1^*) + \sigma\Phi_1^* = \int W(\omega', \omega, r)\sigma_s(r)\Phi_1^*(r, \omega')d\omega' + p(r, -\omega)$$

with the boundary condition

$$\Phi_1^*(r, \omega = 0) \quad \text{if} \quad r \in \Gamma, \quad \omega \cdot n > 0.$$

This is the integro-differential equation with source density

$$p_1(r, \omega) = p(r, -\omega).$$

Now

$$I_p = (\Phi^*, \Phi_0) = \int_R \int_\Omega \Phi^* \Phi_0 dr d\omega = \int_R \int_\Omega \Phi_1^*(r, \omega)\Phi_0(r, -\omega) dr d\omega. \tag{5.11}$$

The relations just obtained show that to estimateI_p, the transfer from a source with density $p(r, -\omega)$ may be simulated and Eq. (5.11), which defines the detector readings with the weight function $\Phi_0(r, -\omega)$, may be calculated. This is the statement of the theorem of optical mutuality.

For practical use of this theorem, Eq. (5.11) must be rewritten as a functional of the collision density,

$$I_p = \int_R \int_\Omega f_1^*(r, \omega) \frac{\Phi_0(r, -\omega)}{\sigma(r)} dr d\omega.$$

Thus, at each collision, the quantity $\Phi_0(r, -\omega)/\sigma(r)$ must be calculated.

5.2. CALCULATIONS OF DERIVATIVS AND PERTURBATIONS

With the help of the weight method problems of transfer theory may be solved simultaneously for several systems by use of only one and the same appropriate Markov chain. This is possible if the kernel functions of the corresponding transfer equations and a given Markov chain satisfy Eq. (1.11). This method is especially effective when a dependence of results on small variations of the parameters of the system is investigated. Such a technique is called the method of dependent sampling.

Suppose that functions h, ψ and the kernels of the integral equation, are functions that depend on some parameter λ:

$$\psi_\lambda(x) = \psi(x, \lambda), \quad k_\lambda(x', x) = k(x', x, \lambda), \quad h_\lambda(x) = h(x, \lambda).$$

Consequently,

$$I(\lambda) = (f_\lambda, h_\lambda) = \sum_{n=1}^{N} Q_n(\lambda) \cdot h(x_n, \lambda). \tag{5.12}$$

In the dependent-sampling algorithm, direct simulation is frequently realilzed for some particular value of the parameter $\lambda = \lambda_0$, i.e.,

$$r(x', x)[1 - p(x')] = k(x', x, \lambda_0), \quad r_0(x) = \psi(x, \lambda_0).$$

After the transition from x' to x a new weight is obtained by

$$Q(\lambda) = Q'(\lambda) \cdot \frac{k(x', x, \lambda)}{k(x', x', \lambda_0)}.$$

In practice, to simulate transitions $x' \to x$, successive simulations of elementary random variables - the direction of scattering, the free-path length, the collision-type number which is sometimes introduced into the phase space - are used. The product of the conditional-probability densities of these random variables determines $k(x', x, \lambda)$ to within a constant factor that does not depend on λ. Thus, we obtain a simple realization of the algorithm under study: we must multiply the particle weight by the ratio of the corresponding probability densities for the values λ and λ_0 after each elementary sampling.

The 'similar-trajectory' method is a special case of technique described. It is sometimes necessary to carry out calculations for systems that are similar in the following sense. The multiplication of a unit vector by λ for one system is equivalent to multiplication of the effective cross section by λ for another system. Assume, for example, that it is desired to estimate the probability of the passage of a particle through homogeneous layer of medium of thickness H. Obviously, calculations for the thickness λH are equivalent to calculations for ther thickness H with cross section $\sigma \cdot \lambda$ (this follows, for example, from the formula for free-path-length simulation $l = -\ln \alpha / \sigma$).

We have, for the similar-trajectories method,

$$\frac{k(x', x, \lambda)}{k(x', x, \lambda_0)} = \frac{\lambda}{\lambda_0} \exp[-\sigma(\lambda - \lambda_0) l].$$

If the particle escapes from the medium its weight is multiplied by

$$\exp\{-\sigma \Delta l (\lambda - \lambda_0)\},$$

where Δl is the length of the last run in the medium.

The corresponding algorithm can be regarded as a realization of the vectorial estimate of the solution of a system of equations with a diagonal kernel. A simple study of the standard weighted modifications of modeling of a transport process show that the bounds Eq. (2.7) for them are actually close if the medium has a large optical dimension. Thus optimization of the weighted methods in transport theory is appropriately realized by means of the solution of the problem

$$\inf_p \sup_\sigma \int_X \frac{k^2(x, x'; \sigma)}{p(x, x')} dx', \quad x \in X.$$

Solution of this problem is simplified when the kernel k depends only on the difference of the arguments, which is approximately fulfilled if the medium is sufficiently close to being homogeneous. Let us note that the corresponding optimization of the "method of similar trajectories" leads to the following simple problem:

$$\min_p \max_{\sigma_1 \geq \sigma \geq \sigma_2} \int_0^\infty \frac{\sigma^2 \exp(-2\sigma l)}{p(l)} dl, \tag{5.13}$$

where $p(l)$ is the simulating density function of l. Detailed numerical investigations (including also modeling the transport process) showed that here the density

$$p(l) = [\sigma_1 \exp(-\sigma_1 l) + \sigma_2 \exp(-\sigma_2 l)]/2.$$

is practically almost optimal [40]. Let us note that in the case of one equation direct modeling turns out to be optimal in the sense of Eq. (2.9).

According to the Section 2.1, the variances of the weighted estimates are here finite for $\forall \sigma \in [\sigma_1, \sigma_2]$, if $q^2 M < 1$, where M is quantity Eq. (5.13), and $q = \sigma_1/\sigma$.

Suppose that it is possible to differentiate the expression Eq. (5.12) under the expectation sign; then

$$\frac{dI(\lambda)}{d\lambda} = \mathrm{E} \sum_{n=0}^{N} \left[\frac{d \ln Q_n(\lambda)}{d\lambda} + \frac{d \ln h(x_n, \lambda)}{d\lambda} \right] Q_n(\lambda) h(x_n, \lambda).$$

The interchange of the averaging and differentiation is permissible here if there exists a function $\psi(x_n)$ such that

$$\left| \frac{d \ln Q_n(\lambda)}{d\lambda} + \frac{d \ln h(x_n, \lambda)}{d\lambda} \right| \cdot |Q_n(\lambda) h(x_n, \lambda)| \leq \psi(x_n)| \cdot |Q_n(\lambda_0) h(x_n, \lambda_0)|,$$

for $\lambda_0 - \varepsilon < \lambda < \lambda_0 + \varepsilon$ and if

$$\mathrm{E} \sum_{n=0}^{N} |Q_n(\lambda_0) h(x_n, \lambda_0)| \psi(x_n) < +\infty.$$

Notice that the unequality holds, if, for example, $\psi(x_n) < c \cdot n$ and $\| K_1 \| < 1$.

Consider now the vector approach (see Section 2.2), which provides wider possibilities. The kernel of the general one-group integral equation of the transfer theory for the collision 'importance' $\varphi(x)$ in the phase space $X = (R \times \Omega)$ has the form (see Section 5.1).

$$k(x, x') = \frac{q(r)}{2\pi |r' - r|^2} g\left(\omega, \frac{r' - r}{|r' - r|}\right) \sigma(r') e^{-\tau(r', r)} \delta\left(\omega' - \frac{r' - r}{|r' - r|}\right).$$

Here, σ is the total cross-section, g is the scattering phase function, $\tau(r, r')$ is the optical length of the segment (r, r'), $q(r')$ is the survival probability of the radiation quantum undergoing a collision; the δ-function shows that the spatial point r' lies on the ray $r(t) = r + t\omega, t \geq 0$. Suppose now that $\sigma_s(r)$ takes constant values σ_1 and σ_2 on D_1 and D_2respectively,where D_1 and D_2 are two disjoint spatial domains. Using the expression for the optical length

$$\tau(r,r') = \int_0^{|r'-r|} \sigma(r(t))dt$$

we obtain

$$\frac{\partial k(x,x')}{\partial \sigma_i} = k(x,x')b_i, \quad b_i = \left\{\frac{\delta_i(r)}{\sigma_i} - \frac{\delta_i(r)}{\sigma(r)} + \frac{\delta_i(r')}{\sigma(r')} - \Delta_i l\right\}$$

$$\frac{\partial^2 k(x,x')}{\partial \sigma_i \partial \sigma_j} = k(x,x')\left\{b_i b_j - \frac{\delta_i(r)\delta_j(r)}{\sigma_i \sigma_j} + \frac{\delta_i(r)\delta_j(r)}{\sigma^2(r)} - \frac{\delta_i(r')\delta_j(r')}{\sigma^2(r')}\right\}, \tag{5.14}$$

where $i,j = 1,2$; $\delta_i(r)$ is the indication function of the domain D_i; $\Delta_i l$ is the length of the part of the ray $r(t)$ lying in D_i. The inequality $\rho(K) < 1$ in the case of the transfer equation considered follows from the inequality $\sigma_s(r)/\sigma(r) \leq q < 1$ or from the boundedness of the medium which leads to the fact that the particle escapes the medium with probability 1 (see, for example, [28]). If one of these conditions holds, then the integral operators with the kernels of type Eq. (5.14) are bounded and $\rho(K) < q_0 < 1$ if $\sigma_i < c < +\infty$. Besides, for all values of σ_1' and σ_2' lying inside the intervals $\sigma_1 \pm \varepsilon$, $\sigma_2 \pm \varepsilon$, respectively, with ε being sufficiently small, the following inequality holds:

$$k(x,x') \leq (1 + c_0\varepsilon)k_\varepsilon(x,x'),$$

where $k_\varepsilon(x,x')$ is the kernel of the transfer equation corresponding to $\sigma_i' = \sigma_i - \varepsilon$, $\quad i = 1,2$. By Theorem 2.4, it is thus possible to use the triangular systems of integral equations considered above to construct the Monte Carlo estimates for the derivatives of the solution with respect to σ_1 and σ_2. Theorem 2.3 shows that the variances of such estimates are finite, for example, if the direct simulation of the transfer process is carried out, i.e. when $p(x,x') = k(x,x')$. It is then not difficult to see that the estimate of the first derivative coincides with that obtained by the differentiation of the estimate for the solution (see, for example, [28]). However, the vector technique permits one to obtain the most general condition under which the estimates are unbiased and have finite variances. Besides, the technique provides a possibility for constructing and investigating estimates for higher order derivatives.

Consider now the functional convergence of the estimates $\xi_x(\sigma_s)$ for the function $\varphi(x) = \varphi(x,\sigma_s)$, where σ_s is the scattering cross-section in a domain D where $\sigma_s^{(1)} \leq \sigma_s \leq \sigma_s^{(2)}$. It is obvious that

$$\frac{k(x,y,\sigma_s)}{k(x,y,\sigma_s^{(1)})} \leq \begin{cases} \sigma_s^{(2)}/\sigma_s^{(1)}, & y \in D. \\ 1, & y \notin D. \end{cases}$$

Consequently, we may put

$$k_0(x,y) = \frac{\sigma_s^{(2)}}{\sigma_s^{(1)}} k(x,y,\sigma_s^{(1)}).$$

By Theorem 2.5 (taking into account the corresponding remark), the functional convergence exists if $p(x,y)$ is chosen such that $\mathrm{D}\xi_{x,0} < +\infty$. This is the case, for example,if the direct simulation without absorption for the operator K_0 is carried out and the following inequallity holds:

$$\frac{\sigma_s^2}{\sigma_s^{(1)}} \sup q(r) < 1.$$

The conditions obtained are essentially extended if the domain D is not large. Moreover, the corresponding extension can be obtained using geometric arguments.

Note that the calculation of the derivatives with respect to $\sigma_c = \sigma - \sigma_s$, the absorption coefficient in D is simpler as compared to the case of the scattering coefficient. For example, if the direct simulation is carried out without absorption in the domain D [i.e. $p(x, y)$ is obatined from $k(x, y)$ by replacing σ with σ_s in d], then the weight in the trajectory points not lying in D is calculated as $\exp(-L\sigma_c)$, where L is the trajectory length inside the domain D (see, for example, [28]). The variances of the estimates for the derivatives are then finite and the functional convergence holds if $\sigma_c \geq 0$.

Note that vector Monte Carlo algorithms (see [28]) are applied to solve systems of transfer equations taking into account polarization as well as the multi-group system of equations of transfer. The technique presented is obviously generalized on these cases by using block-triangular systems of integral equations which are obtained by differentiating original systems.

Consider now an example of evaluation of perturbations in the transfer theory by the Monte Carlo method.

Suppose that a kernel $k_1(x, x')$ is obtained from the kernel $k(x, x')$ by replacing the cross-section $\sigma(r)$ with zero in a spatial domain D. Then

$$\delta k(x, x') = \begin{cases} -k(x, x'), & x' \in D, \\ 0, & x' \notin D. \end{cases}$$

Consequently, from Eq. (2.27) for $\delta h \equiv 0$ we have

$$\varphi_1(x) = \mathrm{E}\eta_x, \quad \eta_x = -\sum_{n=1}^{N} Q_n h(x_n) \sum_{m=0}^{n-1} \delta_D(x_{m+1}), \tag{5.15}$$

where $\delta_D(x)$ is the indication function of the domain D. For the direct simulation [i.e. when $p(x, y) \equiv k(x, y)$] we have

$$\eta_x = -\sum_{n=1}^{N} h(x_n) \sum_{m=0}^{n-1} \delta_D(x_m).$$

A change in the order of summation in Eq. (5.15) yields

$$\eta_x = -\sum_{m=1}^{N} \delta_D(x_m) \sum_{n=m}^{N} Q_n h(x_n).$$

Using the iterated averaging and the collision estimate technique, we can show (see [31]) that

$$\mathrm{E}\eta_x = -\int_\Omega \int_D \varphi^*(x')\varphi(x')dx',$$

where $\varphi^*(x')$ is the collision density.

It is interesting to note that the calculation technique according to expression Eq. (5.15) essentially differs from the Monte Carlo algorithm (see the first description of it in [30]) based on the known formula of the small perturbation theory for the integro-differential transfer equation. In [30], it was actually noted that to realize accurately the formula of the perturbation theory, it is necessary to randomize the standard estimates by simulating additional particles; the algorithm is presented at the end of this Section.

An important property of the algorithm presented consists in the above-mentioned possibility of decreasing the error of the perturbation calculations on the basis of equation (2.24).

In the example considered, the matrix operations are particularly simple because they are applied only to matrices of the type $\Delta(1,-1)$ and $\Delta(1,0) = \{\delta_{ij}\}$ (see Section 2.2). Condition Eq. (2.26) can be verified on the basis of the inequality

$$\rho[R(K)\delta K] \leq \| \delta K \| (1- \| K \|)^{-1}.$$

In the transfer theory, the quantities $\| K \|$ and $\| \delta K \|$ are sufficiently easy estimated using the medium constants (see, for example, [28]).

Note that the transition density $p(x, x')$ must be chosen such that the probability of occurring at least one collision in the perturbation region of D will not be too small. If the domain D has a small volume but a sufficiently large extension in space, then intersections of D without a collision occur often, i.e. collisions inside D occur rarely. It is then convenient to put

$$p(x, x') = k(x, x', t)$$

where $k(x, x', t)$ is obtained from $k(x, x')$ by replacing in D the quantity σ with σt, where $t > 1$. The cost of the corresponding weight algorithm is minimized by an appropriate choice of t.

Now let us consider the general method of realization of the standard perturbation formula for the integro-differential transfer equation with using additional trajectories [30]. With the aim of the description simplicity the method is presented for the case of small perturbations. We introduce the notation:

$$J_1(\delta\sigma) = \int_X \Phi(x)\delta\sigma(x)\Phi^*(x)dx,$$

$$J_2(\delta w_s) = \int_X \int_X \Phi(x)\delta w_s(r, \omega \to \omega')\Phi^*(x')dxdx',$$

where

$$w_s(r, \omega \to \omega') = \sigma_s(r)W(\omega, \omega', r) = \sigma_s(r)g(\mu, r).$$

It is clear that J_2 is a deviation of the number of collisions with the weight Φ^*, while J_2 is a deviation of the number of scattered particles with the similar weight. The standard small perturbation formula has the form [28]:

$$\delta J = -J_1(\delta\sigma) + J_2(\delta w_s),$$

where J is the functional under the consideration.

Consider a space perturbation related to a small parameter h, so that a particle path at the perturbed region is expressed by the formula:

$$\delta l(x) = \alpha(x)h + o(h),$$

provided a particle crosses the boundary of the perturbed region at the phase point x. The corresponding collision probability is equal to

$$\alpha(x)h\sigma(x) + o_1(h).$$

For estimating the number of particles, scattered in the perturbed region, a 'forced' collision in this region is sampled with the weight $\alpha h\sigma$. Necessary values of the function Φ^* can be here estimated statistically by the way of simulating the additional trajectories. Therefore, we use here the 'dual randomization' (see Section 1.6).

Let us formulate now the algorithm in detail. When a trajectory finds itself at the boundary of the perturbed region, the following auxiliary operations are realized:

1) in the nonperturbed medium, a trajectory is constructed with the weight - $\alpha h\delta\sigma$, where $\delta\sigma = \sigma - \sigma'$, σ is a new cross-section and σ' is the original cross-section in the perturbed region;

2) beginning with the 'forced' collision in the new medium, a trajectory is constructed with the weight $\alpha\sigma h$;

3) beginning with the 'forced' collision in the original medium, a trajectory is constructed with the weight $-\alpha\sigma' h$.

The sum of all statistical estimates of the values Φ^* with the above-mentioned weights gives the statistical estimate of the value δJ. The strict justification of this algorithm can be constructed using the repeated averaging as in Section 1.6.

It is clear that it is possible to use one and the same original trajectory for estimating $\Phi^*(x)$ according to item (1) at different points x of the trajectory. After changing the order of summation in the corresponding resulting sum we obtain the bilinear estimate [31,40]:

$$\eta_1 = \sum_{n=0}^{N} \sum_{k=1}^{k_n} (\alpha h \delta\sigma)_k,$$

so that $\mathrm{E}\eta_1 = J(\delta\sigma)$. Here the internal sum is taken over all the intersections (of the trajectory with the above-mentioned boundary) preceding the collision point x_n.

In [41], the algorithms for calculating the derivatives of the coefficient of utilization of the heat neutrons θ for a complicated reactor sell including absorbers in the form of cylinders are given. The coefficient θ is represented in the form

$$\theta = \mathrm{M}h(r_N, i_N).$$

Here r_N is the coordinate of the last collision point, i_N is the corresponding index of the collision which is considered to be an additional coordinate of the phase space. The case $i = 1$ corresponds to the capture: $h(r,i) = 1$, if $i = 1$, and $h(r,i) = 0$ if $i \neq 1$. Now we derive the algorithm for calculating the derivatives with respect to the radius and to the shift of the absorber. To do this, we use the formula of small perturbations. The quantity $\delta\sigma_c$ is increasing far more rapidly than $\delta\sigma_s$ when the radius or the position of the absorber is varying (e.g., 10-50 times in the case of heat neutrons). Therefore, to approximate evaluation of $\delta\theta$, it is sufficient to take only the first term in the formula of small pertuabations. Note that the evaluation can be improved by replacing $\delta\sigma$ with $\delta\sigma_c$, i.e.,

$$\delta\theta \approx -(\Phi\delta_c, \Phi^*) = \delta\theta_0.$$

The calculations made on the basis of one group correlated estimates for a typical cell show that the error of this method is less than 10%. We now describe the algorithms for calculating the biliniar functions of the form $\delta\theta_0$.

Algorithm A (Calculation of the Derivatives of θ with Respect to the Radius of the Absorber). Denote the integral $\delta\theta_0$ correcponding to the increase of the radius of the absorber δr trough $(\cdot)_{\delta r}$. It is known that the evaluation of the quantity $\delta r \delta\sigma_c |\mu_k|^{-1}$ can be carried out by summing the values $\delta r \delta\sigma_c |\mu_k|^{-1}$, where μ_k is the cosine of the angle between the neutron direction and the normal vector to the surface of the absorber at the point of k-th intersection. To evaluate the statistical estimate of the quantity Φ, the subsequent part of the same trajectory can be used. Thus we get following representation

$$(\Phi\delta_c, \Phi^*)_{\delta r} = \mathrm{M}\left\{\left(\sum_k \frac{\delta r \delta\sigma_c}{|\mu_k|}\right) h(r_N, i_N)\right\},$$

where the sum is taken over all intersections of the trajectory with the surface of the absorber. To construct the estimate for the derivative, it is sufficient to devide this expression by δr.

Algorithm B (Calculation of the Derivatives of θ with Respect to H, the Shift of the Absorber). Analogous arguments show that the following representation holds

$$(\Phi\delta_c, \Phi^*)_{\delta H} = \mathrm{M}\left\{\left(\sum_k \frac{\delta\sigma_c \delta H v_k}{|\mu_k|}\right) h(r_N, i_N)\right\},$$

where v_k is the cosine of the angle between the direction of the shift and the normal vector to the surface of the absorber at the point of k-the intersection. Numerical results obtained by the algorithms A and B are reported in [41].

5.3. MULTIVELOCITY RADIATION TRANSPORT PROCESS WITH FISSION

A conventional mathematical model for a stationary radiation-transport process is the integro-differential transfer equation (see, for example, [11,49])

$$\begin{aligned} \omega\nabla\Phi + \sigma\Phi = & \int_V \sigma_s(\mathbf{r}, v')w_s(\mathbf{v}, \mathbf{v}'; \mathbf{r})\Phi(\mathbf{r}, \mathbf{v}')d\mathbf{v}' \\ & + \int_V \nu(\mathbf{r}, v')\sigma_i(\mathbf{r}, v')w_f(\mathbf{v}, \mathbf{v}'; \mathbf{r})\Phi(\mathbf{r}, \mathbf{v}')d\mathbf{v}' + f_0(\mathbf{r}, \mathbf{v}) \\ & \mathbf{r} \in D \subset R_3. \end{aligned} \tag{5.16}$$

Here $\Phi = \Phi(\mathbf{r}, \mathbf{v})$ is the radiation intensity or the density of the flow of appropriate particles; V is the space of the velocities $\mathbf{v} = v \cdot \omega$; $v = |\mathbf{v}|$; $\sigma = \sigma(\mathbf{r}, v)$ is the total cross-section, $\sigma_s(\mathbf{r}, v)$ is the scattering cross-section, $\sigma_f(\mathbf{r}, v)$ is the fission (multiplication) cross-section, with

$$\sigma = \sigma_s + \sigma_f + \sigma_c,$$

where σ_c is the capture (disappearance of a particle) cross-section; $w_f(\mathbf{v}, \mathbf{v}'; \mathbf{r})$ is the indicatrix of fission; $\nu(\mathbf{r}, v')$ is the average number of secondary particles per a fission event initiated by a particle travelling at a speed v' at a point $\mathbf{r}$; and $f_0(\mathbf{r}, \mathbf{v})$ is the distribution density of the particle source.

With Monte Carlo methods, the natural boundary conditions for equation (5.16) are fulfilled automatically (see, for example, [11,49]), however, if we assume that the medium is bounded by a convex surface and that outside the surface $\sigma = \sigma_c \neq 0$, then equation (5.16) can be extended to the entire space [28].

The radiation transport process can be considered to be a homogeneous Markov chain of collisions of a particle with elements of matter. A collision may result in the capture, scattering, or fission of the particle, with the probabilities being σ_c/σ, σ_s/σ, and σ_f/σ, respectively. The particle path terminates at the instant the particle is captured or escape from the medium. If a multiplication takes place, we deal with a branching physical chain. The free path l of the particle between two successive collisions is distributed with the density

$$f_l(t) = \sigma(\mathbf{r}(t), v) \exp\left\{ - \int_0^l \sigma(\mathbf{r}(t_1), v) \mathrm{d}t_1 \right\},$$

where $\mathbf{r}(t) = \mathbf{r}_0 + t\omega$, $\mathbf{r}_0$ is the initial point and ω is the path direction. This density is normalized provided that equation (5.16) is extended to the entire space in the above manner. The collision density $\varphi(\mathbf{r}, \mathbf{v})$ is related [11] to the density of the particle flux by

$$\varphi(\mathbf{r}, \mathbf{v}) = \sigma(\mathbf{r}, v)\Phi(\mathbf{r}, \mathbf{v}). \tag{5.17}$$

The Monte Carlo method, i.e. the numerical simulation of the particle paths, can be used to estimate, for example, the following transport-process functionals. The integral

$$\int_{D_i} \int_V \varphi(\mathbf{r}, \mathbf{v}) \mathrm{d}\mathbf{r}\, \mathrm{d}\mathbf{v} \tag{5.18}$$

is the average number of collisions in a range D_i and estimated by the appropriate sample mean value. If the weight $1/\sigma(\mathbf{r}, v)$ is assigned to the collisions, then by virtue of (5.17) we can estimate the integral at

$$J_i = J_{D_i} = \int_{D_i} \int_V \Phi(\mathbf{r}, \mathbf{v}) \mathrm{d}\mathbf{r}\, \mathrm{d}\mathbf{v}. \tag{5.19}$$

The estimate thus constructed is a special case of the so-called collision estimate.

It is also known (see, for example, [53]) that integral (5.19) equals the mean free path length L_i a particle travels in the range D_i, i.e. $J_i = \mathrm{E}(L_i)$. The corresponding Monte Carlo estimate is referred to as the free-path estimate.

Sometimes it is reasonable to transform equation (5.16) to an equivalent form by introducing fictitious 'delta-scattering'. Suppose that $\sigma(\mathbf{r}, v) \leq \sigma_m$. Add the equality

$$(\sigma_m - \sigma)\Phi = \int_V [\sigma_m - \sigma(\mathbf{r}, v')]\delta(\mathbf{v}' - \mathbf{v})\Phi(\mathbf{r}, \mathbf{v}')\mathrm{d}\mathbf{v}'$$

to equation (5.16). The resultant equation can be treated as a transfer equation in a fictitious medium for which σ_m is the total cross-section, σ is the cross-section of physical collision, and $(\sigma_m - \sigma)$ is the cross-section of fictitious scattering with no change of velocity. In this case σ_m need not be constant. The straightforward simulation of the transport process thus described provides a correct simulation of the free path between physical collisions [39]. In some cases, such a transformation significantly simplifies the simulation of the free path in geometrically complex media (see, for example, [41]). In the sequel, we will show that the use of delta-scattering also can simplify the calculation of parametric derivatives.

The collision density $\varphi(\mathbf{r},\mathbf{v})$ satisfies (see, for example, [28]) the integral equation

$$\varphi(x) = \int_X k(x'x)\varphi(x')\mathrm{d}x' + f(x), \tag{5.20}$$

where $x = (\mathbf{r},\mathbf{v}) \in R_3 \times V$, f is the distribution density for primary collisions and $k(x',x)$ is the distribution density for the average number of secondary collisions that result from a single collision of a particle with a velocity $\mathbf{v}'$ at a point $\mathbf{r}'$, $\mathbf{x}' = (\mathbf{r}',\mathbf{v}')$. Similar to [28], it is easy to derive

$$\begin{aligned} k(x',x) &= \left[\frac{\sigma_s(\mathbf{r}',v')}{\sigma_s(\mathbf{r}',v')} w_s(\mathbf{v},\mathbf{v}';\mathbf{r}') + \nu(\mathbf{r}',v')\frac{\sigma_f(\mathbf{r}',v')}{\sigma_s(\mathbf{r}',v')} w_f(\mathbf{v},\mathbf{v}';\mathbf{r}')\right] \\ &\quad \times \frac{\sigma(\mathbf{r},v)\exp\{-\tau(\mathbf{r}',\mathbf{r};v)\}}{|\mathbf{r}-\mathbf{r}'|^2}\delta\left(\omega - \frac{\mathbf{r}-\mathbf{r}'}{|\mathbf{r}-\mathbf{r}'|}\right), \end{aligned}$$

where

$$\tau(\mathbf{r}',\mathbf{r};v) = \int_0^{|\mathbf{r}'-\mathbf{r}|} \sigma(\mathbf{r}'+t\omega)\mathrm{d}t.$$

This expression is already connected with the weight modification of the process, namely one secondary particle with the weight $\nu(\mathbf{r}',v')$is considered in the case of fission. In a more general weight algorithm (see, for example, [40]) the paths are simulated according to a certain transition density $p(x',x)$, with the auxiliary weight Q being multiplied by the weight factor $k(x',x)/p(x',x)$ after the transition $x' \to x$. However, in the multivelocity transport theory the straightforward implementation of this algorithm is not possible due to incompatible singularities in the indicatrices of different types of scattering, because the particle energy upon scattering may be in a determinate way associated with the direction of scattering according to the scattering type. Therefore in order to construct a general weight algorithm it is necessary 'to stratify' the general distribution of collisions according to their types by introducing the original number k of the collision type into the set of the coordinates in the phase space, i.e. put $x = (k,\mathbf{r},\mathbf{v})$. In this case the new value of x is defined after the collision type has been chosen according to the probabilities σ_k/σ, that is, when modelling a transition $x' \to x$ one should determine the new coordinate in the following succession: $\mathbf{v}$, $\mathbf{r}, k$.

The corresponding modified kernel of equation (5.20) takes the form:

$$k(x',x) = w_{k'}(\mathbf{v},\mathbf{v}';\mathbf{r})\frac{\sigma(\mathbf{r},v)\exp\{-\tau(\mathbf{r}',\mathbf{r};v)\}}{|\mathbf{r}-\mathbf{r}'|^2}\frac{\sigma_k(\mathbf{r},v)}{\sigma(\mathbf{r},v)}\delta\left(\omega - \frac{\mathbf{r}-\mathbf{r}'}{|\mathbf{r}-\mathbf{r}'|}\right). \tag{5.21}$$

Here we may put $\sigma_1 \equiv \sigma_f$ and $w_1 \equiv \nu \cdot w_f$ and take into account various scattering types, including fictitious delta-scattering.

The weight Monte Carlo algorithms are constructed by representing a solution to equation (5.20) by a Neumann series (see, for example, [28]). The availability of n_0 such that $\|K^{n_0}\| < 1$ will suffice for the series to converge. The operator norm is easily estimated in this case if the operator is considered to act from L_1 into L_1. The norm in L_1 is constructed with allowance made for the discreteness of the coordinate k.

In this case we have

$$\|K\| \leq \sup\left(\frac{\sigma_s}{\sigma} + \nu\frac{\sigma_f}{\sigma}\right) = q.$$

Thus, if the branching is not involved in the simulation, in order for the Neumann series to converge it is logical to demand that the condition $q < 1$ be met.

The Monte Carlo method is commonly used to estimate linear functionals of the form:

$$J_n = (\varphi, h) = \int_X \varphi(x)h(x)\mathrm{d}x.$$

For instance, to estimate integral (5.18) put $h(x) = 1$, while to estimate integral (5.19) put $h(x) = 1/\sigma(\mathbf{r}, v)$ for $\mathbf{r} \in D_i$ and $h(x) = 0$ for $\mathbf{r} \notin D_i$. If $\{x_n\}$ is the physical chain of collisions (such a thing is possible within the framework of the model involved, provided $\nu \equiv 1$), then we have

$$J_n = \mathrm{E}\xi, \quad \xi = \sum_{n=0}^{N} h(x_n).$$

If a nonphysical chain of collisions is simulated with the transition density $p(x', x)$, then the collision estimate ξ is of the form:

$$\xi = \sum_{n=0}^{N} Q_n h(x_n),$$

where Q_n are auxiliary weights calculated by the relations:

$$Q_0 = \frac{f(x_0)}{\pi(x_0)}, \quad Q_n = Q_{n-1}\frac{k(x_{n-1}, x_n)}{p(x_{n-1}, x_n)}. \tag{5.22}$$

Here $\pi(x)$ is the initial-state density being modelled. In this case the transition density $p(x', x)$ should include delta functions appearing in the appropriate indicatrices of scattering. The delta functions can be formally eliminated in the ratio $k(x', x)/p(x', x)$ and then the relation $\mathrm{E}\xi = J_n$ is justified in the regular way [28]. The weights are to be considered in more detail in Section 5.4 with the aim to construct the estimates of parametric derivatives of the functionals like J_n.

For the variance of the random quantity ξ we have (see Section 1.3)

$$\mathrm{D}\xi = (\chi, h[2\varphi^* - h]) - (f, \varphi^*)^2,$$

where φ^* is the solution of the adjoint equation $\varphi^* = K^*\varphi^* + h$ and χ is the Neumann series for the equation

$$\chi(x) = \int_X \frac{k^2(x', x)}{p(x', x)}\chi(x')\mathrm{d}x' + \frac{f^2(x)}{\pi(x)}. \tag{5.23}$$

Thus, a simple modification of a physical branching transport process is the Markov chain for which the transition density $p_0(x', x)$ is derived from (5.21) by replacing $\nu \to 1$. Then Q_n is the product of the values of ν for the fission events preceding x_n, while the kernel of equation (5.23) is obtained from (5.21) by changing $\nu \to \nu^2$. Hence the variance of such a simple weight algorithm is finite if the following condition is met:

$$\sup\left(\frac{\sigma_s}{\sigma}+\nu^2\frac{\sigma_f}{\sigma}\right)=q_1<1. \tag{5.24}$$

It should be mentioned that an expression for the variance of the collision estimate, which is constructed on a branching Markov chain, is obtained in [4] (see also [40]). It follows from the expression that in the case of direct simulation of a physical branching process the variance is finite and the value of J_n is also finite, i.e. for a subcritical system.

Now let us construct a weight modification of the free-path estimate [40]. To this end, we replace the chain $\{X_n\}$, as it is done in [31], by the equivalent chain

$$\{y_{n+1}=(x_n,x_{n+1})\}_{n=0,1,\dots}$$

in the phase space Y of the points $y=(x',x)$. Evidently, the transition density $y=(x',x)\to y_1=(x_1',x_1)$ is defined for the new chain as follows:

$$k_1(y,y_1)=k(x,x_1)\delta(x'-x).$$

On the basis of this expression, a weight modification of the free-path estimate is constructed in the ordinary way, provided that the simulated transition density $p(y,y_1)$ also contains a factor $\delta(x_1'-x)$ which can be cancelled in the ratio $k_1(y,y_1)/p_1(y,y_1)$. It follows from (5.27) that here the weight factors remain standard ones, i.e. $k_1/p_1\equiv k/p$, and the weight free-path estimate for integral (5.19) is of the form:

$$\xi_l=\sum_{n=1}^{N}Q_n l_n^{(i)},\qquad y_n=(x_{n-1},x_n),$$

where Q_n are defined by formulae (5.22), while $l_n^{(i)}$ is the path length in D_i from x_{n-1} to x_n. The same argument shows that here equation (5.22) remains practically unchanged and condition (5.24) remains valid. The free-path estimate may prove to be especially effective in calculating the parametric derivatives of the functionals of the type of (5.19) if the range D_i and the range in which the parameter is varied (for example, the total cross-section) are small.

In conclusion, we note that restrictions imposed by condition (5.24) can be weakened if we increase the quantity σ_i by the term $\sigma_c=\sigma-\sigma_f-\sigma_s$ and recalculate the value of ν by the formula:

$$\nu'=\frac{\nu\sigma_f}{\sigma_f+\sigma_c}=\frac{\nu\sigma_f}{\sigma_a}\le\nu.$$

In this case kernel (5.21) and the function φ remain unchanged, while the paths in a finite system terminate with the probability of unity, due to the particle escape.

5.4. CALCULATION THE DERIVATIVES WITH RESPECT TO CROSS-SECTIONS

In $L_\infty(X)$ we consider the following integral equation of the second kind with a kernel dependent on the parameter λ:

$$\varphi^*(x,\lambda)=\int_X k(x,y,\lambda)\varphi^*(y,\lambda)dy+h(x,\lambda) \tag{5.25}$$

which, in the operator notation, is of the form: $\varphi^*=K^*\varphi^*+h$. Here X is a n-dimensional Euclidean space. We pose the problem on a calculation of the derivatives:

$$\varphi^{*(n)}(x,\lambda) = \frac{\partial^n \varphi^*(x,\lambda)}{\partial \lambda^n}.$$

Let $K^{*(n)}$ denote an integral operator with the kernel $k^{(n)}(x,y,\lambda)$ and $\rho(K^*)$ stand for the spectral radius of the operator K^*.

Differentiating (so far only formally) relation (5.25) m times with respect to λ, we arrive at the triangular system of integral equations:

$$\varphi^{*(n)} = \sum_{i=0}^{n} C_n^i K^{*(n-i)} \varphi^{*(i)} + h^{(n)}, \quad n = 0,1,...,m \tag{5.26}$$

or $\mathbf{\Phi}^* = \mathbf{K}^* \mathbf{\Phi}^* + H$.

Vector Monte Carlo estimates have been considered in Section 2.2 for the solution of system (5.26). It has been show that such estimates have finite variances if operators $K_p^{(n)}$ with the kernels $k^{(n)2}(x,y,\lambda)/p(x,y)$, $n = 1,...,m$, are bounded and the spectral radius of the operator K_p with the kernel $k^2(x,y,\lambda)/p(x,y)$ is less than unity. The above conditions agree with the standard conditions ensuring the finiteness of the variances of the scalar Monte Carlo estimates for equation (5.25). In Section 2.2 it also has been noted that the vector estimates of parametric derivatives are equivalent to the scalar estimates obtained by the straightforward differentiation of the collision estimates for the functionals involved. Thus, on the basis of the theory of vector estimates, a simple condition has been obtained for the finiteness of the variances of the scalar estimates of parametric derivatives which will be considered later.

Equations (5.25) is the conjugate of (5.30), but the results of this section are extended to the problems in Section 5.3, because the following representations are valid (see, for example, [39,40]):

$$\xi = \frac{f(x_0)}{\pi(x_0)} \xi_{x_0}, \quad \mathrm{E}\xi_x = \varphi^*(x), \quad J_h = (f, \varphi^*), \quad \mathrm{E}\xi^2 = \mathrm{E}\Big[\frac{f^2}{\pi^2}\mathrm{E}(\xi_x^2 | x = x_0)\Big].$$

In this section we will consider one of the cross-sections σ_k, σ in a certain subrange as a parameter λ.

As noted in Section 5.3, for the transfer equation involved, the relation $\rho(K) < 1$ holds provided $q < 1$ or if the medium is bounded which results in a particle escaping from the medium with the unit probability. Under these conditions integral operators with the kernels of the form of $k^{(n)}(x,y,\lambda)$ are bounded and $\rho(K) < q_0 < 1$ if $\sigma_k \le C < +\infty$, $\sigma_0 = \sigma$. Moreover, for all σ_k' belonging respectively to the interval $\sigma_k \pm \varepsilon$, where ε is a sufficiently small negative number, the following inequality holds:

$$k(x,x') \le (1 + C_0 \varepsilon) k_\varepsilon(x,x'),$$

where $k_\varepsilon(x,x')$ is the kernel of the transfer equation corresponding to $\sigma_k' - \varepsilon$, $k = 0,1,\ldots$. Thus, by virtue of Theorem 2.4 in order to construct the estimates of the derivatives of the solution with respect to σ_k we can use the above triangular systems of integral equations. The variances of such estimates are finite if, for example, the simplest modification of direct simulation is performed (see Section 5.3) and condition (5.24) is met.

Now we will construct the scalar versions of the estimates of the derivatives of the functionals of the form of (5.18), (5.19) with respect to σ, σ_k, and σ_c.

First we estimate the derivative with respect to the total cross-section. If we assume σ_k to be constant, then we will actually estimate the derivative with respect to a scale factor ρ which is introduced for a certain subrange D_j by replacing σ with $\rho\sigma$ and σ_k with $\rho\sigma_k$. The paths are simulated for $\rho = \rho_0$. In this case the weight factor, according to (5.21), is of the form:

$$\frac{k(x'x)}{p_0(x'x)} = \nu_{k'} \left(\frac{\rho}{\rho_0}\right)^{\delta_j(r)} \exp\{-(\rho - \rho_0)\tau_j(\mathbf{r}', \mathbf{r}; v)\}, \tag{5.27}$$

where $\nu_1 \equiv \nu$ and $\nu_k \equiv 1$ for $k \neq 1$; $\delta_j(r)$ is the indicator of the range D_j; $\tau_j(\mathbf{r}', \mathbf{r}, v)$ is the optical path length in D_j from $\mathbf{r}'$ to $\mathbf{r}$ at $\rho = 1$. For definiteness, we will consider the derivative of functional (5.19), putting $h(x) \equiv \delta_i(r)$. From the above it follows that the estimate ξ can be differentiated here term by term, i.e. constructing the estimate for the quantity $J_j^{(m)}$ amounts to finding the expression for $Q^{(m)}$, where $Q \equiv Q_n$ is the current weight. On the basis of the representation

$$Q' = Q(\ln Q)'$$

we obtain

$$Q^{(m)} = \sum_{k=1}^{m} u_k Q^{(m-k)} \frac{(m-1)!}{(k-1)(m-k)!} \,. \tag{5.28}$$

It is obvious that

$$u_1 = (\ln Q)' = \frac{m_j}{\rho} - T_j, \quad u_k = (\ln Q)^{(k)} = \frac{(-1)^{k-1}}{\rho^k} m_j (k-1)! \,. \tag{5.29}$$

Here m_j is the total number of preceding collisions (including the current collision) in D_j and T_j is a similar total optical path length in D_j for $\rho = 1$. Because we have to calculate sum (5.28), the highest derivatives can be estimated using the modification like the capture estimate (see, for example [40]). The estimate is calculated in this case only at capture, but with the additional weight σ/σ_c. Because the number of the collision type is considered to be an extra coordinate in the phase space, this estimate can be treated here as a special version of the collision estimate.

Now let us consider the estimate of the derivative with respect to the cross-section σ_k of some scattering type (or fission with fixed ν) and assume this quantity to be constant in D_j. Since the quantity $\sigma(\mathbf{r}, v)$ is cancelled in (5.21), the quantities u_k are also determined by formulae (5.29) with the following substitutions:

$$\rho \to \sigma_k, \quad \rho_0 \to \sigma_k^{(0)}, \quad T_j = L_j,$$

where L_j is the total length of the previous path of a particle in D_j and m_j is the total number of the preceding scattering events of the k type in D_j.

Now let us estimate the derivative with respect to ρ in case auxiliary fictitious delta-scattering is used (see Section 5.3). We assume that $\rho\sigma \leq \sigma_m$; the appropriate weight factor of the form:

$$\frac{k(x', x)}{p_0(x', x)} = \nu_{k'} \cdot \left(\frac{\rho}{\rho_0}\right)^{\delta_j(\mathbf{r})}$$

for a physical collision and

$$\frac{k(x',x)}{p_0(x',x)} = \nu_{k'} \cdot \left[\frac{\sigma_m}{\sigma_m - \sigma\rho_0}\left(1 - \frac{\sigma}{\sigma_m}\rho\right)\right]^{\delta_j(\mathbf{r})}$$

for delta-scattering. The estimate by formulae (5.28) can be obtained here in the ordinary way; however, at σ/σ_m = const inD_j we also can calculate Q' immediately by differentiating the product

$$\rho^{m_j}\left(1 - \frac{\sigma}{\sigma_m}\rho\right)^{n_j}, \tag{5.30}$$

where m_j is the total number of physical collisions and n_j is the total number of delta-scattering events in the range D_j.

In calculating the derivatives with respect to σ_k, the weight factors are of the same form on substitution:

$$\sigma\rho \longrightarrow \sigma_k, \quad \sigma\rho_0 \longrightarrow \sigma_k^{(0)}$$

and if σ_k/σ_m = const in D_j, the quantity Q' can be calculated by differentiating the product

$$\sigma_k^{m_j}\left(1 - \frac{\sigma}{\sigma_m}\rho\right)^{n_j},$$

where m_j is now the total number of k-type scatterings in D_j. The expressions obtained should be used if auxiliary delta-scattering is employed to equalize cross-sections (see, for example, [41]).

Let us mention a possible effective application of delta-scattering to stochastic problems. Assume that ρ is a random field and for simplicity assume that σ is a constant. One path of a particle is constructed for each realization ρ in estimating $\mathrm{E}\Phi_\rho$, the mean value of the solution when a double-randomization method is used. The weight is given by a more general expression than (5.30):

$$\left(\prod_{i=1}^{m_j}\rho_i\right)\prod_{i=1}^{n_j}\left(1 - \frac{\sigma}{\sigma_m}\rho\right)^{n_j}.$$

The appropriate collision estimate can be averaged (for a fixed path) over the filed distribution ρ. Evidently, such averaging reduces to calculating power moments and requires mush less information on the field ρ than its direct modelling, which, in addition, is not always possible. A similar procedure can be laid down for estimating other moments $\mathrm{E}\Phi_\rho^m$.

Let us assume now that a constant s is added to the total cross-section (and thereby to the capture cross-section $\sigma_c = \sigma - \sum_k \sigma_k$) in the range D_j. We find from (5.21) that

$$\frac{k(x',x)}{p_0(x',x)} = \nu_{k'} \cdot \exp\{-l_j s\},$$

where l_j is the path length in D_j for the corresponding free path. Let $\tilde{\nu}$ denote the product of the previous values of ν_k. Then we obtain

$$Q = \tilde{\nu}\exp\{-L_j s\},$$

where L_j is the total length of the previous path in D_j. Hence we have

$$Q^{(m)} = \tilde{\nu}(-L_j s)^m.$$

5.5. CALCULATING CRITICAL VALUES OF THE PARAMETERS: THE CRITICAL DENSITY, THE TIME CONSTANT OF PARTICLE MULTIPLICATION, THE EFFECTIVE MULTIPLICATION FACTOR

Let us write the integro-differential particle-transport equation in the form:

$$L\Phi + \sigma\Phi = S\Phi + \lambda S_f\Phi + f_0, \tag{5.31}$$

where (see [11,49] and Section 5.3) $L\Phi \equiv \omega\nabla\Phi$.

$$S\Phi \equiv \int_V \sigma_s(\mathbf{r}, v')w_s(\mathbf{v}, \mathbf{v}'; \mathbf{r})\Phi(\mathbf{r}, \mathbf{v}')d\mathbf{v}'$$

$$S_f\Phi \equiv \int_V \nu(\mathbf{r}, v')\sigma_f(\mathbf{r}, v')w_s(\mathbf{v}, \mathbf{v}'; \mathbf{r})\Phi(\mathbf{r}, \mathbf{v}')d\mathbf{v}'.$$

The principal (smallest in magnitude) eigenvalue $\lambda = \lambda_0$ is known to be positive under rather wide assumptions and can be found by the standard iteration method [11,49]. The quantity $1/\lambda_0 = k_{\text{eff}} = 1$ is called the effective multiplication factor of a system. If $k_{\text{eff}} < 1$ the system is subcritical; if $k_{\text{eff}} > 1$ it is supercritical; and at $k_{\text{eff}} = 1$ it is critical. Taking into account that Φ depends regularly on λ (see [11,49]), we differentiate (5.31) m times with respect to λ to obtain

$$L\Phi^{(m)} + \sigma\Phi^{(m)} = S\Phi^{(m)} + \lambda S_f\Phi^{(m)} mS_f\Phi^{(m-1)} \tag{5.32}$$

or $R\Phi^{(m)} = mS_f\Phi^{(m-1)}$. Hence

$$\Phi^{(m)} = m!(R^{-1}S_f)^{(m-1)}\Phi'.$$

It has been shown that the operator $R^{-1}S_f$ has spectral properties [10] sufficient for the Kellog iterations to converge. Therefore we have

$$m\Phi^{(m-1)}/\Phi^{(m)} \to \lambda_1, \quad m \to \infty$$

and, in the limit, equation (5.32) takes the form:

$$L\Phi^* + \sigma\Phi^* = S\Phi^* + (\lambda + \lambda_1)S_f\Phi^*$$

i.e., $k_{\text{eff}} = (\lambda + \lambda_1)^{-1}$. Similarly, for linear functionals J of Φ of the form of (5.18), (5.19) we have

$$mJ^{(m-1)}/J^{(m)} \to \lambda_0 - \lambda = k_{\text{eff}}^{-1}(1 - \lambda k_{\text{eff}}).$$

Here the estimates of the quantities $J^{(m)}$ are made and justified [39,40] by straightforward differentiation of the weight collision estimate (see section 5.3)

$$\xi^{(m)} = \sum_{n=0}^{N}(\lambda^{m_n})^{(m)}Q_nh(x_n),$$

where m_n is the total number of the previous fission events. The estimate ξ is constructed on a modified path with $\lambda\nu \equiv 1$. The finiteness of the appropriate variances is ensured by the relation (see section 5.3):

$$q_0 = \sup\left(\frac{\sigma_s}{\sigma} + \lambda^2\nu^2\frac{\sigma_f}{\sigma}\right) < 1.$$

Therefore it is reasonable to make equivalent modifications of the process by substituting

$$\sigma_f \to \sigma_f + \sigma_c, \quad \lambda\nu \to \frac{\lambda\nu\sigma_f}{\sigma_f + \sigma_c} < \lambda\nu.$$

Using the results of [4], we can show that the direct modelling of a branching physical process can be used to construct the estimates of $I^{(m)}$ with finite variances at $\lambda < k_{\text{eff}}^{-1}$. The algorithm of estimating k_{eff} has a number of advantages over the standard method of 'generations with a fixed number of points' (see Section 5.7). The estimates constructed with the weighted method or a correlated sampling can be realized by using the same ensemble of paths for different values of the system parameters. The estimate of the functional relations under study is thus effectively improved [20]. Using the standard procedure (Section 5.3 and [37]), we also can estimate the corresponding parametric derivatives and the perturbations of the quantities $I^{(m)}$ and, consequently, we can estimate k_{eff}. In the method of generations, such estimates are very hard to obtain (see Section 5.7).

Now let us consider the problem of estimating a time constant τ^* of particle multiplication, which determines the time exponential asymptotics $\exp(-\tau^* t)$ of the function Φ (see, for example, [11,49]). It is known [11,49] that if the quantity τ^*/v (v is the particle speed) is added to the absorption cross-section, the system becomes critical, i.e. τ^* is an eigenvalue of the equation that is derived from (5.31) by replacing

$$\sigma \to \sigma + \frac{\tau}{v}.$$

After the equation thus modified is differentiated m times, we obtain

$$L\Phi^{(m)} + \left(\sigma + \frac{\tau}{v}\right)\Phi^{(m)} = S\Phi^{(m)} + S_f\Phi^{(m)} - \frac{m}{v}\Phi^{(m-1)}$$

or $R_1\Phi^{(m)} = -(m/v)\Phi^{(m-1)}$. Hence

$$\Phi^{(m)} = (-1)^{(m-1)} m!(R_1^{-1}v)^{-(m-1)}\Phi'$$

and

$$m\Phi^{(m-1)}/\Phi^{(m)} \to \tau_1, \quad m \to \infty.$$

In the limit we have

$$L\Phi^* + \left(\sigma + \frac{\tau}{v}\right)\Phi^* = S\Phi^* + S_f\Phi^* - \frac{\tau_1}{v}\Phi^*$$

and $\tau^* = \tau + \tau_1$, i.e. we have here

$$mJ^{(m-1)}/J^{(m)} \to \tau^* - \tau.$$

It is easy to obtain the expression which is quite well known [40]:

$$Q_n = \tilde{\nu}\exp(-\tau t_n),$$

where t_n is the total path time of a particle to r_n and $\tilde{\nu}_n$ is the weight, which takes into account fission events if the simplest modification of the process without branching is realized.

This implies that

$$Q_m^{(m)} = \tilde{\nu}_n(-t_n)^m \exp(-\tau t_n),$$

i.e. if $\tau > 0$ the quantity τ_1 is negative and *vice versa.* Since the quantity τ/v is added to the capture cross-section, then the quantity q_0 modified by replacing $\sigma_f \to \sigma_f + \sigma_c$ can be made less than unity by choosing sufficiently large τ. It is very easy to justify here [39,40] the unbiasedness of the estimates of the derivatives and the finiteness of their variances at $q_0 < 1$ on the basis of the vector approach [39,40]. It has been mentioned above that the condition of the finiteness of the variances is less restrictive in modelling the branching. In addition to the above advantages over the method of generations, we now get another advantage: we need not interpolate by the values of $k_{\text{eff}}(\tau)$. The interpolation has serious drawbacks in this case (see Section 5.7).

Now let us consider the estimate of the critical value of a scale factor ρ by which all the cross-sections in a certain range D_j are multiplied (see section 5.4). On substituting

$$\sigma \to \rho\sigma, \quad \sigma_s \to \rho\sigma_s, \quad \sigma_f \to \rho\sigma_f$$

for $r \in D_j$ in equation (5.31) and on differentiating this equation m times with respect to ρ, we get

$$L\Phi^{(m)} + \rho T\Phi^{(m)} = -mT\Phi^{(m-1)},$$

where $T = \sigma - S - S_f$. Repeating the arguments of the preceding examples, we obtain the following limit equation:

$$L\Phi^* + (\rho + \rho_1)\sigma\Phi^* - (\rho + \rho_1)S\Phi^* + (\rho + \rho_1)S_f\Phi^*,$$

where $\rho_1 = \lim(m\Phi^{(m-1)}/\Phi^{(m)})$.

Hence we have here

$$mJ^{(m-1)}/J^{(m)} \to \rho^* - \rho.$$

Monte Carlo estimates for the quantities $I^{(m)}$ have been considered in Section 5.3. Although the quantity ρ has no effect on q_0, the variances of the collision estimates and hence the variances of the appropriate estimates of the quantities $I^{(m)}$ can be made finite [39,40] by choosing a sufficiently small value of ρ, because a particle can disappear due to its escape from the system.

Similar to the above example, we can replace here the function Φ by a function $\varphi = \sigma\Phi$ because

$$\frac{m\varphi^{(m-1)}}{\varphi^{(m)}} = \frac{m[(m-1)\Phi^{(m-2)} + \rho\Phi^{(m-1)}]}{m\Phi^{(m-1)} + \rho\Phi^{(m)}} \underset{m\to\infty}{\longrightarrow} \rho_1$$

and hence we can use functional (5.18) to construct computational estimates.

5.6. NUMERICAL EXAMPLES

Consider a single-velocity particle transport process in a ball of radius R with isotropic scattering and with the constants:

$$\sigma = 1, \quad \sigma_f = 0.03, \quad \sigma_s = 0.97, \quad \sigma_c = 0, \quad \nu = 2.5.$$

A diffusion approximation with the extrapolated boundary is used as a mathematical model of the process (see, for example, [49]):

$$\Delta\Phi + c\Phi = -g, \quad \Phi\big|_{R+\alpha/\sigma} \equiv 0 \tag{5.33}$$

$$c = 3(\nu - 1)\sigma_f\sigma_s = 0.13095, \quad \alpha = 0.71044\ldots .$$

Here $g = g(\mathbf{r})$ is the volume distribution density of an isotropic particle source and $r = |\mathbf{r}|$ is the distance from the ball centre. Within the limits of approximation (5.33), the critical value R is given by the formula [49]:

$$R_{\mathrm{cr}} = \frac{\pi}{\sqrt{3(\nu - 1)\sigma_f\sigma_s}} = \frac{\alpha}{\sigma} = 7.971107.$$

1. Now we introduce a scale factor ρ by substituting

$$\sigma \to \rho\sigma, \quad \sigma_s \to \rho\sigma_s, \quad \sigma_f \to \rho\sigma_f. \tag{5.34}$$

It is logical to take a value of ρ that ensures the finiteness of the variances of the appropriate weight Monte Carlo estimates as an initial value for which we will calculate the parametric derivatives. In other words, we should take ρ such that the system is subcritical on substitution $\nu \to \nu^2$ (see section 5.3). For such substitution we have

$$R_{\mathrm{cr}}^{(1)} = \frac{\pi}{\sqrt{3(\nu^2 - 1)\sigma_f\sigma_s}} - \frac{\alpha}{\sigma} = 3.930042.$$

Hence the value of ρ suitable in the above sense must satisfy the inequality:

$$\rho < R_{\mathrm{cr}}^{(1)}/R_{\mathrm{cr}} = 0.493036\ldots .$$

For this reason we carried out the numerical experiment in the diffusion approximation for a ball with radius R_{cr} and with the constants like those in (5.34) at $\rho = 0.493036$.

Using Green's function for the Helmholtz equation at $g(\mathbf{r}) = \delta(\mathbf{r}' - \mathbf{r})$ we obtain from (5.33):

$$\Phi(r) = \frac{\sin[a\rho + b - r\rho\sqrt{c}]}{4\pi r \sin(a\rho + b)}, \quad a = R_{\mathrm{cr}}\sqrt{c}, \quad b = \frac{\alpha}{\sigma}\sqrt{c}. \tag{5.35}$$

This quantity can be treated as the total intensity at the ball centre from a 'unit-power' source situated at a distance of r from the centre.

To simplify further arguments we put $g(\mathbf{r}) = 1/r$ and as an original functional we consider the following integral (with allowances made for the Jacobian r^2)

$$J(\rho) = \rho\int_0^{R_{\mathrm{cr}}} r\Phi(r)\mathrm{d}r = C[f_1(\rho) + f_2(\rho)]$$

$$f_1(\rho) = \frac{\cos(b)}{\sin(a\rho + b)}, \quad f_2(\rho) = -\frac{\cos(a\rho + b)}{\sin(a\rho + b)}.$$

The derivatives $f_2^{(m)}(\rho)$ are easy to calculate:

$$f_2'(\rho) = \frac{a}{\sin(a\rho + b)} = a[1 + f_2^2(\rho)], \quad f_2^{(m)} = a(f_2^2)^{(m-1)}, \quad m > 1.$$

The derivatives of the function $1/\sin(a\rho + b)$ were calculated as follows:

$$f(\rho) = \frac{1}{\sin(a\rho + b)}, \quad f_1(\rho) = f'(\rho) = -\frac{a\cos(a\rho + b)}{\sin^2(a\rho + b)}$$

$$f'' = 2f_1^2 f^{-1} + a^2 f, \quad f''' = 6f_1^3 f^{-2} + 5a^2 f_1$$

$$f^{(4)} = 24 f_1^4 f^{-3} + 28a^2 f_1^2 f^{-1} + 5a^4 f$$
$$f^{(5)} = 120 f_1^5 f^{-4} + 180a^2 f_1^3 f^{-2} + 61a^4 f_1$$
$$f^{(6)} = 720 f_1^6 f^{-5} + 1320a^2 f_1^4 f^{-3} + 662a^4 f_1^2 f^{-1} + 61a^6 f$$
$$f^{(7)} = 5040 f_1^7 f^{-6} + 10920a^2 f_1^5 f^{-4} + 7266a^4 f_1^3 f^{-2} + 1385a^6 f_1.$$

The results of the calculations are summarized in Table 5.1, where $\tilde{\rho}_m = mI^{(m-1)}/I^{(m)} + 0.493036$.

Table 5.1. The results of the solution of the model problem

m	$f_2^{(m)}$	$J^{(m)}$	$\tilde{\rho}_m$
1	2.91870	3.2242	
2	1.83337	10.11978	1.1302
3	50.2969	63.1890	0.9735
4	124.2049	483.533	1.0158
5	3523.80	4859.84	0.9905
6	17990.03	56675.4	1.0075
7	532951	792139	0.9939

The results obtained confirm that $\tilde{\rho}_m \to \rho_{cr} = 1$ according to a power law. Note that this example is rather representative because the diffusion approximation especially well takes account of the contribution of multiple scattering. Here scattering increases in importance in the estimates of $I^{(m)}$ as $m \to \infty$. It is quite obvious that the convergence rate of $\tilde{\rho}_m \to \rho_{cr}$ can be increased by choosing some good approximation to Φ^* for g.

Note that one can try to obtain similar results on the basis of the Cauchy-Hadamard formula (see the end of Section 4.1), but the corresponding convergence may be very slow. For instance, instead of the values $\{\tilde{\rho}_m\}$ from Table 5.1 the Cauchy-Hadamard formula gives the following values (beginning with $m = 1$): 0.803, 0.938, 0.949, 0.965, 0.970, 0.976, 0.979.

2. The system considered above for $R = R_{cr}$ is critical. As noticed in Section 5.5 the system remains critical if one add the quantity τ^*/v to the absorption cross section. Thus, $\tau^* = 0$ in our case. Let us replace σ by $(\sigma + \tau/v)$. Then the value $\tau^* = 0$ will be the eigenvalue of the corresponding equation (5.31).

Let us take as an initial one the value of τ which guarantees finiteness of the variances of weighted estimates. The parametric derivatives will be calculated at this point. Criterian (5.24) implies

$$\frac{\sigma_s}{\sigma + \tau/v} + \nu^2 \frac{\sigma_f}{\sigma + \tau/v} < 1, \qquad \frac{\tau}{v} > \sigma_s + \nu^2\sigma_f - \sigma = 0.1575.$$

Therefore, the numerical experiment was carried out for $\tau_0 = 0.16$ and the ball radius equal to R_{cr}.

With the Green function $G(r, r')$ for the equation (5.33) we obtain for $g(r) = \delta(r' - r)$

$$\tilde{\Phi}(r) = G(r,0) = \frac{\sin(a+b-r\sqrt{c})}{4\pi r \sin(a+b)},$$

where

$$a = R\sqrt{c} = 2.8845, \qquad b = \frac{\alpha}{\sigma+\tau/v}\sqrt{c}, \quad c = 3[\sigma_f(\nu-1) - \frac{\tau}{v}]\sigma_s.$$

The quantity $\tilde{\Phi}$ means the total particle flux measured at the centre of a ball which is induced by the "unit pover" source placed at the distance of r from the centre.

Put

$$g(r) = \begin{cases} 1/r, & R_0 < r < R, \\ 0, & \text{otherwise}, \end{cases} \tag{5.36}$$

where R_0 is a constant small as compared to R. Then the solution of problem (5.33) for $r = 0$ takes the form

$$\Phi(0) = \int_{\mathbf{D}} \tilde{\Phi}(r) g(r)\, dr.$$

Consider the following integral as an initial functional (taking into account the Jacobian r^2) :

$$J(\tau) = \left(\sigma + \frac{\tau}{v}\right)\Phi(0) = \left(\sigma + \frac{\tau}{v}\right)\int_0^R \tilde{\Phi}(r) r^2 g(r)\, dr = P(gf_1 + f_2),$$

where P is a constant, $g = e^{2\alpha u\sqrt{3\sigma_s}}$, $u = \frac{\sqrt{\gamma}}{H+\gamma}$, $\gamma = \frac{\tau}{v} - (\nu-1)\sigma_f$, $H = \sigma + \sigma_f(\nu - 1)$,

$$f_1 = f(A) = \frac{e^{F\sqrt{\gamma}}(e^{A\sqrt{\gamma}} - 1)}{u(ge^{2F\sqrt{\gamma}} - 1)}, \quad f_2 = f(-A),$$

$$A = (R - R_0)\sqrt{3\sigma_s}, \quad F = r\sqrt{3\sigma_s}.$$

The derivatices of the functions f_1, u, g with respect to τ are equal to

$$u' = u\Big(\frac{1}{2\gamma} - \frac{1}{H+\gamma}\Big), \quad g' = 2Bu'g,$$

$$f_1' = \Big(\frac{F+A}{2\sqrt{\gamma}} - \frac{u'}{u}\Big)f + \frac{e^{F\sqrt{\gamma}}}{ge^{2F\sqrt{\gamma}} - 1}\Big[\frac{A}{2u\sqrt{\gamma}} - f\Big(g' + g\frac{F}{\sqrt{\gamma}}\Big)e^{F\sqrt{\gamma}}\Big],$$

$$J'(\alpha) = P(g'f_1 + gf_1' + f_2').$$

The results of computations for $R_0 = R/5$ and $R_0 = R/3$ (see Table 5.2) confirm the convergence

$$\tau_m^* = \frac{mJ^{(m-1)}}{J^{(m)}} + \tau_0 \longrightarrow \tau^* = 0 \quad \text{as} \quad m \to \infty.$$

Now let us consider calculations of $\boldsymbol{\tau}^*$ based on the simulation of trajectories. We study here the transfer process, which is characterized by the same constants and R ($R = 7.971107$) and the same distribution density of the particle source (5.36).

Consider as an initial functional the following inner product:

$$J = \Phi(0) = (\Phi(r), \delta(r)).$$

It is convenient for this problem to use the adjoint transfer equation (see Section 5.1). By the theorem of optical mutuality we find

Table 5.2. The results of τ^* computations [26] based on the diffusion approximation

m	$R_0 = R/5$		$R_0 = R/3$	
	$J^{(m)}/P$	τ^*_m	$J^{(m)}/P$	τ^*_m
1	−8.98	0.0141	−6.28	0.0508
2	114.8	0.0035	89.4	0.0196
3	−2172	0.0015	−1173	0.0087
4	54543	0.0007	45499	0.0042

$$J = (\Phi(r), \delta(r)) = (\Phi^*(r), g(r)),$$

where $\Phi^*(r)$ is the solution of the adjoint equation

$$-L\Phi^* + \left(\sigma + \frac{\tau}{v}\right)\Phi^* = S\Phi^* + S_f\Phi^* + \delta(r).$$

Thus, to find the functional J one can simulate the transfer process from the source with density $\delta(r)$ and determine the reading of detector with the weight function $g(r)$.

It was remarked in Section 5.4 that to estimate $J^{(m)}$ one can differentiate the collision estimate. For the nonphysical collision chain with the initial density

$$\pi(x_0) = \frac{(\sigma + \tau_0)\exp\left(-(\sigma + \tau_0)r_0\right)}{r_0^2}$$

and transition density

$$p(\vec{x}_1, \vec{x}) = \frac{\sigma_k}{\sigma + \tau_0} w_{k_1}(\vec{v}, \vec{v}_1, \vec{r}) \frac{(\sigma + \tau_0)\exp\left(-(\sigma + \tau_0)|\vec{r} - \vec{r}_1|\right)}{|\vec{r} - \vec{r}_1|^2} \delta\left(\vec{\omega} - \frac{\vec{r} - \vec{r}_1}{|\vec{r} - \vec{r}_1|}\right)$$

the estimate of the functional J equals

$$\xi = \sum_{n=0}^{N} Q_n\, g(x_n).$$

Here $x = (\vec{r}, \vec{v}, k)$, k is the number of the collision type, Q_n are auxiliary random weights calculated by the formulae

$$Q_0(\tau) = \frac{f(x_0)}{(\sigma + \tau)\pi(x_0)}, \qquad Q_n(\tau) = \frac{1}{\sigma + \tau} Q_{n-1}(\tau) \frac{k(x_{n-1}, x_n)}{p(x_{n-1}, x_n)},$$

$f(x)$ is the density of first collisions,

$$f(x_0) = \frac{(\sigma + \tau_0)\exp\left(-(\sigma + \tau_0)r_0\right)}{r_0^2},$$

$k(x_1, x)$ is the distribution density of the average number of secondary collisions followed the collision of the particle having the velocity $\vec{v}_1$ at the point $\vec{r}_1$:

$$k(x_1, x) = \frac{\sigma_k}{\sigma} \tilde{\nu}_1\, w_{k_1}(\vec{v}, \vec{v}_1, \vec{r}) \frac{\sigma \exp\left(-(\sigma + \tau)|\vec{r} - \vec{r}_1|\right)}{|\vec{r} - \vec{r}_1|^2} \delta\left(\vec{\omega} - \frac{\vec{r} - \vec{r}_1}{|\vec{r} - \vec{r}_1|}\right).$$

The quantity $\tilde{\nu}_1$ takes value 2.5, when the fission occurs, otherwise $\tilde{\nu}_1 = 1$.

The following representations for the weights can be obtained by simple transformation:

$$Q_0(\tau) = \frac{\exp(-(\tau - \tau_0)r_0)}{\sigma + \tau_0}, \qquad Q_n(\tau) = \tilde{\nu}_n \frac{\exp(-(\tau - \tau_0)t_n)}{\sigma + \tau_0},$$

where $\tilde{\nu}_n = \nu^{m_n}$, m_n is the number of fissions up to the n-th collision of particle, t_n is total path time of the particle up to the n-th collision.

Differentiation of the weights m times yields

$$Q_0^{(m)}(\tau_0) = \frac{r_0^m\,(-1)^m}{\sigma + \tau_0}, \qquad Q_n^{(m)}(\tau_0) = \tilde{\nu}_n \frac{(-1)^m\,t_n^m}{\sigma + \tau_0}.$$

The variance of the weight estimate can be decreased by simulating the process without absorption, but with the replacements

$$\sigma_f \to \sigma_f + \sigma_c\,, \qquad \nu \to \frac{\nu\sigma_f}{\sigma_f + \sigma_c}. \tag{5.37}$$

As the result of this replacement the trajectories (and computation time) grow about 7 times longer. The data presented in Table 5.3 show that the approximation errors of $\tilde{\tau}_m^*$ and $\hat{\tau}_m^*$ (20 000 trajectories being used) are in order of magnitude less than those of τ_m^* (100 000 trajectories), which were calculated on the trajectories without the replacement (5.37). The quantities σ_m, $\tilde{\sigma}_m$, $\hat{\sigma}_m$ are the majorants of the mean square errors.

Table 5.3. The results of computations [26] based on the Monte Carlo method

m	$R_0 = R/5$		$R_0 = R/3$
	$\tau_m^* \pm \sigma_m$	$\tilde{\tau}_m^* \pm \tilde{\sigma}_m$	$\hat{\tau}_m^* \pm \hat{\sigma}_m$
1	0.043 ± 0.003	0.044 ± 0.001	0.070 ± 0.001
2	0.011 ± 0.006	0.012 ± 0.002	0.031 ± 0.002
3	0.003 ± 0.011	0.005 ± 0.003	0.015 ± 0.002
4	0.001 ± 0.017	0.003 ± 0.004	0.007 ± 0.004
5	-0.006 ± 0.025	0.001 ± 0.005	0.004 ± 0.005
6	-0.010 ± 0.036	-0.0003 ± 0.006	0.001 ± 0.006
7	-0.015 ± 0.050	-0.002 ± 0.008	-0.001 ± 0.008
8	-0.019 ± 0.068	-0.004 ± 0.010	-0.003 ± 0.010
9	-0.024 ± 0.086	-0.006 ± 0.012	-0.005 ± 0.012
10	-0.030 ± 0.107	-0.007 ± 0.015	-0.006 ± 0.015

The convergence τ^* to 0 is better, when $g(r)$ is a more exact approximation for Φ^* (see Table 5.3, where value $R_0 = R/5$ corresponds to the better approximation).

Notice that we observe some negative drift of the estimate τ_m^* for sufficiently large m. Similar phenomenon appeared also in Section 4.1. This effect could be explained by the tendency of understating (in modulus) of high derivatives by the Monte Carlo method.

5.7. MONTE CARLO CALCULATIONS OF CRITICAL SYSTEMS WITH EQUALIZATION OF GENERATIONS

Estimation of critical parameters of nuclear reactors is usually reduced to the computation of the value k_{eff}, i.e., the coefficient of neutron multiplication per one generation.

The computational method, considered below, can be determined as 'the method of generations with a fixed number of fission points' [17,29,57]. Denote:

x – a phase point of a fission; i.e., a space point and a velocity of a neutron, resulting a fission;

X – the phase space;

X^n – the set-theoretic production $X \times X \times ... \times X$;

$\overset{n}{x}= (x^{(1)}, x^{(2)}, ..., x^{(n)})$ – an element of the space X^n.

The coefficient k_{eff} is the first eigenvalue of the operator L, determined in the space of bounded variation functions by the formula

$$[L_p](\cdot) = \int_X l(x, \cdot)p(dx),$$

where $l(x, B)$ is the average number of fissions in the region B that result from a single fission at the point x. The quantity

$$k(x) = l(x, X)$$

stands for the local multiplication coefficient per one generation with respect to the point x.

If $p^*(\cdot)$ is the first eigenmeasure of the operator L, then the following equation is valid:

$$k_{\text{eff}}p^*(\cdot) = \int_X l(x, \cdot)p^*(dx).$$

It is assumed that $p^*(X) = 1$, $p^* \geq 0$. If an algorithm of neutron trajectories simulation is determined, then the random variable k is determined, so that

$$\mathrm{E}k_x = k(x).$$

In the case of the direct simulation k_x is a random number of fission in X that result from one fission at the point x.

Let us now formulate the algorithm of the method of generations with a fixed number of fission points [17,29,57]. Let

$$\overset{n}{x}_0= \left(x_0^{(1)}, x_0^{(2)}, ..., x_0^{(n)}\right)$$

be fixed. Beginning with the points $\{x_0^{(i)}\}$ trajectories of fission neutrons are constructed up to terminations, i.e., fission, capture or escape from the system. Here we obtain n_1 new fissions:

$$\overset{n_1}{x}_1= \left(x_0^{(1)}, x_0^{(2)}, ..., x_0^{(n_1)}\right).$$

Simultaneously there is computed a value of the random variable

$$k_{\overset{n}{x}} = \frac{1}{n}\sum_{i=1}^{n} k_{x_0^{(i)}}.$$

A new generation $\overset{n}{x}_1$ is constructed from $\overset{n_1}{x}_1$ in the following way: if $n_1 > n$, then $n_1 - n$ points are randomly rejected, while if $n_1 < n$, then $n - n_1$ points are sampled with replacement from $\overset{n_1}{x}_1$ or $\overset{n}{x}_0$ and added to $\overset{n_1}{x}_1$ [17,29,57]. As a result we have the sequence

$$\overset{n}{x}_0, \overset{n}{x}_1, \ldots, \overset{n}{x}_N$$

which is the homogeneous Markov-chain in X^n. The value k_{eff} is estimated by the formula

$$k_{\text{eff}} \approx k_n^{M,N} = \frac{1}{N-M+1} \sum_{j=M}^{N} k_{\overset{n}{x}_j}.$$

Clearly, it is expedient to choose the value M so, that $k_{\overset{n}{x}_j}$ and $k_{\overset{n}{x}_{j+M}}$ are really independent. Computations show [17,29], that $M = 5$ is usually satisfactory.

Under conditions, usually realized in practice, the following relation is valid [17,19]

$$k_n^{(M,N)} \xrightarrow{P} k_m \approx k_{\text{eff}},$$

and, what is more, the distribution of $k_n^{M,N}$ converges to the Gaussian distribution with the average k_n and the variance σ^2 (about estimating σ see below). Therefore, using the sequence $\{\overset{n}{x}_j\}$ it is possible to construct the confidence interval for k_n. Furhter, it is known [57] that

$$|k_{\text{eff}} - k_n| = O(n^{-1}),$$

$|k_{\text{eff}} - k_{100}| \lesssim 0.001$, and, as a rule, $k_n < k_{\text{eff}}$.

Now let us consider the question about estimating the value σ. The variance per one generation can be estimated by the formula:

$$d_n \approx \tilde{d}_n = \frac{1}{N-M} \sum_{j=M}^{N} (\eta_j - \tilde{\eta})^2.$$

Suppose that for the sequence $\{k_{\overset{n}{x}_j}\}$ the correlation coefficients are small and only the coefficient ρ_1 of the first order essentially differs from zero (this supposition is supported by computations). Then

$$\sigma = \sigma(k_n^{M,N}) \approx \frac{\tilde{\sigma}_n}{\sqrt{N-M+1}} \sqrt{1 + 2\tilde{\rho}_1},$$

where $\tilde{\sigma}_n = \sqrt{\tilde{d}_n}$ and

$$\tilde{\rho}_1 = \frac{N-M+1}{N-M} \frac{\sum_{j=M}^{N-1} (\eta_j - \tilde{\eta})(\eta_{n+1} - \tilde{\eta})}{\sum_{j=M}^{N} (\eta_j - \tilde{\eta})^2}.$$

For the case of the direct simulation, when k_x is the number of the corresponding sampled fissions, it is possible to obtain an approximate expression of $\sigma = \sqrt{d_n}$ through k_n. Consider the value

$$d_n = \bar{\mathrm{D}}\left(\frac{1}{n} \sum_{i=1}^{n} k_{x^{(i)}}\right) \approx \frac{1}{n^2} \sum_{i=1}^{n} \bar{\mathrm{D}} k_{x^{(i)}}.$$

The symbol $\bar{\mathrm{D}}$ means the total variance, while D means the conditional variance with respect to the fixed $x^{(i)}$. The latter equality is approximate because of the dependence between values $k_{x^{(i)}}$, but in practice this approximation is often satisfactory. It is assumed that all $x^{(i)}$ have the same distribution $p_n(\cdot) \approx p^*(\cdot)$ [17,29].

Therefore

$$\bar{\mathrm{D}}(k_{x^{(i)}}) = \bar{\mathrm{D}}(k(x^{(i)})) + \mathrm{ED}k_{x^{(i)}}.$$

According to [17,29] let us suppose that the value $\mathrm{D}(k(x^{(i)}))$ is relatively small, i.e.,

$$\int k^2(x)p_n(dx) \approx \Big(\int k(x)p_n(dx)\Big)^2 = k_n^2. \tag{5.38}$$

Then

$$\mathrm{D}(k_x) \approx \mathrm{ED}(k_x),$$

where x is distributed with the density $p_n(\cdot)$. Let

$$m = \mathrm{entier}(\nu).$$

When simulating the transfer process we construct m or $m+1$ trajectories with respect to the probabilities $m+1-\nu$ and $\nu+m$ from a fission point. Therefore

$$k_x = y_1 + y_2,$$

where y_1 is the result sum of m Bernoully samples with the parameter $p = k(x)/\nu$ (this is related to the Wald identity [19]), while the distribution of the random variable y_2 is determined by the formulas:

$$P(y_2 = 1) = (\nu - m)k(x)/\nu$$
$$P(y_2 = 0) = 1 - (\nu - m)k(x)/\nu.$$

Since random variables y_1 and y_2 are independent, we have

$$\begin{aligned}\mathrm{D}(k_x) = \mathrm{D}(y_1) + \mathrm{D}(y_2) &= m\frac{k(x)}{\nu}\Big(1 - \frac{k(x)}{\nu}\Big) + (\nu - m)\frac{k(x)}{\nu}\Big(1 - (\nu - m)\frac{k(x)}{\nu}\Big)\\ &= k(x) - ((\nu - m)^2 + m)\frac{k^2(x)}{\nu^2}.\end{aligned}$$

Then with allowance for (5.38)

$$\mathrm{ED}k_x \approx k_n - ((\nu - m)^2 + m)\frac{k_n^2}{\nu^2}.$$

Since

$$\frac{(\nu - m)^2 + m}{\nu^2} = -\frac{(\nu - m)(m + 1 - \nu)}{\nu^2} + \frac{1}{\nu},$$

then

$$\mathrm{ED}k_x \approx k_n\Big(1 - \frac{k_n}{\nu} + \frac{(\nu - m)(m + 1 - \nu)}{\nu^2}\Big)$$

As a result

$$\sigma_n = \sqrt{d_n} \approx \sqrt{\frac{k_n(1 - k_n/\nu + (\nu - m)(m + 1 - \nu)/\nu^2)}{n}}.$$

The sufficiency of this approximation is supported by computations [29].

Now let us consider the problem of estimating critical values of parameters of nuclear reactors. Let c be a parameter so that $k_{\mathrm{eff}} = k_{\mathrm{eff}}(c)$. If

$$k_{\text{eff}}(c^*) = 1,$$

then c^* is critical value of c. Further, if

$$k_1 = k(c_1) < 1, \quad k_2 = k_2(c_1) > 1,$$

then

$$c^* \approx \tilde{c}^* = c_1 + \frac{1 - k_1}{k_2 - k_1}(c_2 - c_1). \tag{5.39}$$

Using the linearization method we obtain the approximate expression

$$\sigma(\tilde{c}^*) \approx \frac{|c_2 - c_1|}{(k_2 - k_1)^2}\sqrt{(1 - k_2)^2\bar{\mathrm{D}}(\tilde{k}_1) + (1 - k_1)^2\bar{\mathrm{D}}(\tilde{k}_2)}. \tag{5.40}$$

The random estimates $\tilde{k}_1$, $\tilde{k}_2$ of k_1, k_2 are supposed to be independent.

Formula (5.39) is not totally convenient, because for the close k_1, k_2 formula (5.40) is not sufficient and $\sigma(\tilde{c}^*)$ is large, while for the distant k_1, k_2 formula (5.39) is not satisfactory. Therefore it is desirable to obtain dependent estimates of $k(c)$ for different values c. Sometimes those estimates can be constructed by using a weighted Monte Carlo method within one generation.

Denote: L_c is the operator, corresponding to the parameter value c, L_c^* is the corresponding conjugate operator. Then

$$L_c p_c = k(c) p_c, \quad L_c^* q^*(c) = k(c) q_c^*.$$

For the value $c = c_0$ the index is omitted, for instance, $p \equiv p_{c_0}$. The quantity $k(c)$ can be represented as

$$k(c) = \frac{(q_c^*, L_c p)}{(q_c^*, p)}.$$

Therefore, if the value function q_c^* changes weakly, then

$$k(c) \approx \frac{(\chi, L_c p)}{(\chi, p)}, \quad \chi \equiv 1.$$

When simulating the generation process for $c = c_0$, it is possible to estimate the quantity $(\chi, L_c p)$ with the use of the weighted algorithm (see section 5.4).

Sometimes it is useful to estimate the derivative $k' = dk/dc$. By differentiating the equality

$$L_c p_c = k(c) p_c$$

with respect to c we get

$$k'p + kp' = L'p + Lp',$$

where L' is the operator with the kernel obtained by differentiating the kernel of the operator L_c with respect to c for the value $c = c_0$. Hence,

$$k'(q^*, p) + k(q^*, p) = (q^*, L'p) + (q^*, Lp').$$

However,

$$(q^*, Lp') = (L^*p^*, p') = k(q^*, p^*),$$

then

$$k' = \frac{(q^*, L'p)}{(q^*, p')}.$$

If q^* changes weakly, then

$$k' \approx \frac{(\chi, L'p)}{(\chi, p')}, \quad \chi \equiv 1.$$

This formula is supported by computations and can be essentially useful for estimating the value c^*. Note, that algorithms for estimating the value $(\chi, L'p)$ are presented in Section 5.4.

5.8. SOLVING SOME INVERSE AND STOCHASTIC PROBLEMS OF THE TRANSFER THEORY

Assume that there is a set of measured values $\tilde{J}_k$ of the functionals

$$J_k(\sigma_1, ..., \sigma_n) = (f, h_k) = (\Phi, p_k), \quad k = 1, ..., N,$$

where f is the solution of the integral transfer equation, and Φ is the corresponding intensity (see section 5.1). We want to evaluate parameters $\sigma_1, ...\sigma_n$. If $(\sigma_1^{(0)}, ..., \sigma_n^{(0)})$ are initial estimations of these parameters, then in the linearization approximation we have the system:

$$\sum_{i=1}^{n} a_{ki}\delta_i = \tilde{J}_k, \quad k = 1, ..., N,$$

where $J_k \equiv J_k(\sigma_1^{(0)}, ..., \sigma_n^{(0)})$,

$$a_{ki} = \frac{\partial J_k(\sigma_1^{(0)}, ..., \sigma_n^{(0)})}{\partial \sigma_i}, \quad \delta_i = \sigma_i - \sigma_i^{(0)},$$

or in the vector form:

$$A\delta = \tilde{J} - J, \quad \delta = \sigma - \sigma^{(0)}. \tag{5.41}$$

The derivatives $\{a_{ki}\}$ are calculated by the Monte Carlo method simultaneously with $\{J_k\}$ (see section 5.2). If the linearization is not sufficient, it is possible to use the successive iterations.

System (5.41) is, generally speaking, overdetermined but it can be solved by the least square method, i.e., by minimization of the quadratic form:

$$(A\delta - (\tilde{J}^* - J), \ A\delta - (\tilde{J}^* - J)).$$

As a result we have the system

$$A^*A\delta = A(\tilde{J}^* - J) \tag{5.42}$$

with the quadratic matrix A^*A. Before using the least square method it is expedient to divide the k-th line of the original system (5.41) by the middle-square error s_k of $\tilde{J}_k - J_k$ and to multiply the i-th column of the matrix A by the $\sigma_i^{(0)}$ for $k = 1, ..., N$ and $i = 1, ..., n$. Therefore the values δ_i are divided by $\sigma_i^{(0)}$. In this way instead of (5.41) we obtain the system

$$[A_1^{-1}AA_2](A_2^{-1}\delta) = A_1(\tilde{J} - J),$$

where

$$A_1 = \mathrm{diag}(s_1, ..., s_N), \quad A_2 = \mathrm{diag}(\sigma_1^{(0)}, ..., \sigma_n^{(0)}).$$

Sometimes the additional information is given in the form:

$$\sum_{i=1}^{n} \sigma_i H_i = \tau, \tag{5.43}$$

where values $\{H_i\}$ and τ are known. In this case it is natural to use the conditional least square method with respect to condition (5.43). The corresponding numerical algorithm is described, for instance, in [2].

Now suppose that σ is a random vector with the correlation matrix C and the functionals $J_k(\sigma)$ depend on σ almost linearly. Note that the last assumption can be considered in the neighbourhood of the point $\sigma^{(0)} = \mathrm{E}\sigma$ (the probability of hitting this neighbourhood is close to unity). Then the correlation moments of the vector J and the autocorrelation moments of the vectors σ and J can be calculated by the linearization of $J(\sigma)$ using the Monte Carlo evaluations of the derivatives a_{ki}. In [2], this calculation technique is developed for the case of the piecewise-constant random scattering cross-section; here the linearization approximation appears to be sufficient.

When *a priori* distribution of σ is given, system (5.41) is solved by the statistical regularization method, i.e., it is reduced to the system [56]:

$$(A^* A_1^{-2} A + C^{-1})\delta = A^* A_1^{-2}(\tilde{J} - J).$$

If *a priori* distribution is Gaussian, then the latter system gives the Bayesian, i.e., the best statistical estimates of σ [56].

In [2], there are considered model inverse problems of the restoration of the altitude curve for the aerosol scattering coefficient with the help of the diurnal horizon brightness for the wavelength $\lambda = 0.55\ \mu m$. There is used condition (5.43), where τ is the known atmospheric optical depth and $\{H_i\}$ are depths of the atmospheric slabs, in which cross-sections are supposed to be constant. The 'conjugate walks' method (see Section 5.1) for estimating the values J_k and a_{ki} is used, but the single-scattering contributions in these values are calculated analytically. It appeared that a possible albedo error does not practically influence on results. The allowed variations of an aerosol indicatrix are not essential, if the observation version is so that the first scattering angle is equal to 57°, because aerosol indicatrixes approximately coincide at this angle point [2].

Some stochastic problems of radiation transfer theory are presented below.

As a first example, let us consider the problem of radiation transfer in a medium represented as a random set of spherical inhomogeneities. It is assumed that the midpoints of the spheres can be considered to form a spatial Poisson point flux, i.e. the numbers of midpoints in non-overlapping domains are independent and distributed according to the Poisson law. Intersections of the spheres are allowed. Denote by σ_1 and σ_2 the total cross-sections of the medium inside and outside the spheres, respectively. Assume that $\max(\sigma_1, \sigma_2) = \sigma_1$. Simulation of trajectories for a fixed realization of such a medium can be carried out by the method of maximal cross-section (see Section 5.1). In this method, the free path length is sampled according to the formula $l = -\ln\alpha$, where α is a random variable uniformly distributed on $(0,1)$; if the sampled collision point does not lie in a sphere, then a 'delta-scattering' is simulated, i.e. the particle moves in the same direction with probability $(\sigma_1 - \sigma_2)/\sigma_1$. Sometimes it is useful to apply a weight technique. The variance of this method for the media studied can be estimated as described in [40].

To construct a realization of the medium, it is sufficient to sample N, the number of sphere midpoints, according to the Poisson distribution and then to sample all the midpoints independently and uniformly in the entire domain. To reduce the number of arithmetic operations, it is possible to divide the domain into parts and to sample the midpoints in a part only when the particle hits this part. Clearly, the size of the part must be larger than the radii of the spheres. The sampled portion of the point flux must be stored because the particle may repeatedly hit the corresponding part of the domain. Resampling of the portions of the flux is not in accordance with the randomization principle of Section 1.8, resulting in a bias of the estimate which can be neglected only for strong scattering anisotropy.

Another general model of stochastic medium follows from the representation of the total cross-section $\sigma(r)$ in the form

$$\sigma^{(m)}(r) = \sum_{i=1}^{m} \sigma_i^{(m)}(r), \tag{5.44}$$

where $\sigma_i^{(m)}, i = 1, ..., m$ are independent of a homogeneous random field related to stationary point fluxes as described in Section A.2. Recall that the correlation function of the field σ coincides with the correlation function of each term $\sigma_i^{(m)}$ in Eq. (5.44), while the one-dimensional distribution is determined by the composition rule. Therefore, representation of Eq. (5.44) is suitable if the one-dimensional distribution of the field simulated is infinitely divisible.

Various properties of fields are considered in Section A.2, where we define a class of correlation functions $k(t)$ for isotropic fields, which is close to the class of convex functions. This provides satisfactory models of random fields required to solve many practical problems of transfer theory. It should be noted that the only reliable information about the correlation characteristics of a realistic random field is often the correlation scale (for example, spatial scale) ρ, which is usually defined by the formula

$$\rho = \int_0^{\infty} k(t)dt.$$

However, ρ can be defined by the expression

$$\rho_1 = \left[-\frac{k(0)}{2k''(o)}\right]^{1/2} \tag{5.45}$$

as well.

Note that $\rho = +\infty$ in the case of the isotropic-layer field discussed in Section A.2. Therefore, the correlation scale of this model is determined by Eq. (5.45). The correlation scale ρ for more complicated 'parallelepipedal' models, where the independent point fluxes are sampled on coordinate axes (see Section A.2), is finite: $\rho < +\infty$.

As follows from the arguments of Section A.2, it is useful to take a large value of m in Eq. (5.44) since the finite-dimensional distributions of the field $\sigma(m)$ become absolutely continuous as $m \to \infty$. This property is natural for fields with absolute-continuous one-dimensional distributions. However, in the problems of transfer theory it is possible to choose m such that the mean size of regions where $\sigma^{(m)}$ is constant is essentially less that the mean free path of the particle. It is preferable to obtain such a value of m in preliminary calculations. There are various algorithms simulating the free path of the particle in a medium with total cross-section of form of Eq. (5.44). Perhaps the simplest but not the most effective is here an algorithm where the free path length is sampled independently for each term $\sigma_i^{(m)}(r)$ and then the minimum of the sampled lengths is chosen. The distribution of this quantity coincides with the physical distribution of the free path length.

Some results of Monte Carlo calculations of transfer problems for stochastic plane-parallel media were obtained. It was shown that these results are in good agreement with the asymptotics of the mean intensity of radiation propagating in such media.

These asymptotics show that it is important to take into account the stochastic inhomogeneity of the actual media. The asymptotics also can be used to check the Monte Carlo calculations. In certain cases, a coefficient in the asymptotics can be effectively calculated by the Monte Carlo method. Therefore, we will consider these asymptotics in detail.

It is well known (see, for example, [11]) that the asymptotics of the radiation intensity in a plane parallel layer are determined to within a constant factor by solution of the transfer equation in an infinite medium. Correspondingly, the following model can be introduced.

(1) The medium density is defined as a homogeneous random field with plane symmetry; more exactly, the scattering and absorption coefficients depend on a single coordinate z, namely,

$$\sigma_s = \sigma_s(z) = q\sigma(z), \quad \sigma_c = \sigma_c(z) = (1-q)\sigma(z), \quad 0 \le q < 1$$

and the scattering indicatrix is the same everywhere. The random function $\sigma(z)$ is homogeneous (with respect to z) and satisfies the conditions of the central limit theorem [19] for $\tau(z) = \int_0^z \sigma(z')dz'$.

(2) The source is uniformly distributed on the plane $z = 0$, with the angular distribution of the source intensity satisfying the characteristic equation of transfer theory [11]. Therefore, the radiation intensity for a given realization of σ is fully determined by the optical length measured from the observation point to the plane $z = 0$:

$$I(z,\sigma) = I(\tau(z)) = e^{-\tau/L}, \quad \tau = \int\limits_0^z \sigma(z')dz',$$

where L is the diffusion length (L equals the first eigenvalue of the characteristic equation for $\sigma = 1$, $\sigma_s = q$, i.e. it is dimensionless diffusion length for the given scattering indicatrix and the fixed survival probability of a quantum in the scattering act).

(3) It is necessary to obtain the asymptotics of the mean intensity as $z \to \infty$, i.e. to find the expectation $EI(z,\sigma)$. The formulated problem is thus generalized for an arbitrary statistical moment of the random intensity $EI^k(z,\sigma)$ by replacing L with L/k.

Using the central limit theorem for homogeneous random functions, the author obtained the following asymptotics:

$$\mathrm{E}I(z,\sigma) \approx \left\{ -\frac{z\sigma_0}{L}\left(1 - \frac{S^2\rho}{L\sigma_0}\right)\right\}, \tag{5.46}$$

where $S^2 = \mathrm{D}\sigma$ and ρ is the correlation length of the random function$\sigma(z)$. Asymptotics of Eq. (5.46) are applicable, if $2S^2\rho < L\sigma_0$.

In addition, [40] contains accurate asymptotics of the mean radiation intensity $E\exp\{-\tau/L\}$ for the Poisson model of $\sigma(z)$ with the parameter λ. As shown in [40], approximate asymptotics of Eq. (5.46) agree well with the exact asymptotics, provided that $2S^2\rho < \sigma_0 L$. Note that the asymptotics of the mean intensity can be estimated for more complicated multi-dimensional models of $\sigma(z)$ of form of Eq. (5.44) using the asymptotics derived by multiplication and by averaging the corresponding one-dimensional asymptotics. It is especially simple when carried out for the isotropic-layer field (see Section A.2).

Let us discuss the practical application of the considered asymptotics. Assume that the radiation flux is incident from above on a plane-parallel layer of medium considered as a homogeneous random field. Also assume the scattering inside the layer to be anisotropic that the horizontal migration of the quantum is less than the correlation length of the medium. To estimate the intensity of the transmitted radiation, it is possible to apply the formula obtained under the condition that the averaging of the intensity is carried out over a region with a diameter essentially larger than the horizontal correlation length of the medium. This averaging is equivalent to the averaging over all field realizations provided that the density field is ergodic; hence, the technique considered above is indeed applicable. Thus, one may conclude, for example, that the radiation passing through a stochastic cloud may exceed the radiation passing through a deterministic homogeneous cloud with the asymptotics $c_0\exp(-\sigma_0 z/L)$.

5.9. THE 'FREE-PATH' ESTIMATE FOR SOLVING THE TRANSFER EQUATION IN TOTAL

It is possible to use the free-path estimate (see section 5.3) for global solving the transfer equation. In the case of spatially one-dimensional problem this equation is written in the form:

$$v\frac{\partial\Phi(z,\mu,v)}{\partial z} + \sigma(z,v)\Phi(z,\mu,v)$$
$$= \sigma_1(z,v)\int w(\mu',v',\to \mu,v;z)\Phi(z,\mu',v')d\mu'dv' + \Phi_0(z,\mu,v).$$

Here $\sigma_1(z,v)$ is the scattering cross-section, $w(\mu',v' \to \mu,v;z)$ is the indicatrix of scattering, $\sigma(z,v)$ is the total cross-section, μ is the z-orth, v is the velocity of a particle. The free-path estimate is constructed by virtue of the following fact: the integral, via the slab with the thickness h of the function

$$\varphi(z) = \int_V \int_{-1}^{1} \Phi(z,\mu,v)d\mu dv$$

is equal to the mean time of the particle living in this slab. The unbiased estimate of J_h will be symbolized by $\tilde{J}_h$, i.e.,

$$\mathrm{E}\tilde{J}_h = J_h.$$

The function $\tilde{J}_h/h$ is the estimate of the quantity $\varphi(z)$ with the finite variance (see section 5.3). The set $\{\tilde{J}_h/h\}$ for the given h-partition of the considered slab H determines the functional estimate of the 'histogram'-type; the uniform devition of this estimate from $\varphi(x)$, provided that evident conditions are satisfied has in the average the order h. If values $\{\tilde{J}_h/h\}$ are considered in the centres of the subintervals, then using the continuous and piecewise-linear prolongation we obtain the estimate $\tilde{\varphi}(z)$ of the 'frequency polygon' type; this estimate uniformly deviates from $\varphi(z)$ in the average to the quantity of order h^2, provided $\varphi(z)$ is piecewise smooth and continuous.

Note, that in spite of the fact that the variance of the similar 'collision estimates' has the order h^{-1}, usually for realistic values of h the free-path estimate is not better than the collision estimate in terms of the simple criterion of $\max\{\mathrm{D}(\tilde{J}_h/h)\}$. This is because of the fact that the free-path value in the plane slab is non-bounded and, for instance, in an empty slab, i.e., when

$$\sigma(z) \equiv 0, \quad z_0 \geq z \geq z_0 + h,$$

the variance of the free-path estimate is infinite. However it is clear, that this estimate gives a more regular approximation of $\varphi(z)$ and can be essentially better than the collision estimate in the sense of functional convergence. Section 5.9 is devoted to the investigation of this problem. First of all note, that the quantity

$$\tilde{\varphi}'(z) = \frac{d\tilde{\varphi}(z)}{dz}$$

is evidently some functional collision estimate and the corresponding variance has the order h^{-1}, while the variance of the estimate of the same derivative, when using a collision frequency polygon (see section 1.5), has the order h^{-3}, because of the weak dependence of collision amounts in adjacent subintervals. Further, it is clear, that the question about convergence of $\tilde{\varphi}$ to φ with respect to some metric can be investigated on the basis of the quantity

$$B(\varphi, \tilde{\varphi}) = \mathrm{E}\|\varphi(z) - \tilde{\varphi}(z)\|^2_{L(H)},$$

where $L(H)$ is the corresponding Banah space [3]. It is expedient to connect this equation with the estimate cost, i.e., with the mean amount of operations which provide the inequality

$$B(\varphi, \tilde{\varphi}) < \delta^2.$$

For the space $L_2(H)$ we have

$$B(\varphi, \tilde{\varphi}) = \mathrm{E}\int_H [\varphi(z) - \tilde{\varphi}(z)]^2 dz \geq \int_H \mathrm{D}\tilde{\varphi}(z) dz + \int_H [\varphi(z) - \mathrm{E}\tilde{\varphi}(z)]^2 dz.$$

Hence, provided the second order derivative of the function $\varphi(z)$ is uniformly bounded, the problem of the cost minimization here can be formulated in the form

$$S_0 = nt_0h^{-1} \to \min_{n,h}, \quad \frac{d_0}{n} + c_0h^4 = \delta^2. \tag{5.47}$$

Here n is sample size (i.e., the amount of simulated trajectories),

$$t_0 = t\text{mes}(H),$$

where t is the cost of a trajectory, and the sense of the quantity t_0 is clear. By analogy with Section 1.5, it is possible to obtain the following optimal, in the sense of (5.47), values:

$$h_0^* = (5c_0)^{1/4}\delta^{1/2}, \quad n_0^* = \frac{5}{4}d_0\delta^{-2}, \quad S_0^* = \frac{5}{4}(5c_0)^{1/4}d_0t_0\delta^{-5/2}$$

Here $B(\varphi, \tilde{\varphi}) < \delta^2$ and therefore $\tilde{\varphi}$ converges to φ with respect the metric L_2, provided $\delta \to 0$. When using a collision estimate, the optimal values of S has also the order $\delta^{-5/2}$.

It is possible to obtain the similar results with respect to the metric $C(H)$ on the basis of the imbedding theorem for the case

$$W_2^{(1)}(H) \to C(H),$$

i.e., on the basis of the relation:

$$\|\varphi - \tilde{\varphi}\|_c^2 < K_0\Big[\int_H [\varphi(z) - \tilde{\varphi}(z)]^2 dz + \int_H [\varphi'(z) - \tilde{\varphi}'(z)]^2 dz\Big]^2.$$

As a result we have the following asymptotic form of the cost minimization problem:

$$S_1 = nt_0h^{-1} \to \min_{n,h}, \quad \frac{d_1}{nh} + c_1h^2 = \delta^2 \tag{5.48}$$

Here we take into account the fact that $\tilde{\varphi}'(z)$ is a piecewise-constant function.

The optimal in the sense of (5.48) values are:

$$h_1^* = (2c_1)^{-1/2}\delta, \quad n_1^* = 2d_1(2c_1)^{1/2}\delta^{-3}, \quad S_1^* = 4d_1c_1t_0\delta^{-4}.$$

When using the collision estimate the similar optimal value of S has the order δ^{-5}.

Chapter 6

Solution of nonlinear problems

6.1. SOLUTION OF NONLINEAR INTEGRAL EQUATIONS WITH THE USE OF BRANCHING CHAINS

Let us consider a nonlinear integral equation of the form:

$$\varphi(x) = \int_X k(x, x')\varphi^n(x')dx' + h(x) \tag{6.1}$$

or in the operator form:

$$\varphi = K\varphi^n + h.$$

We consider this equation in $L_\infty(X)$. Let us also consider the parametrized equation

$$\varphi_\lambda = \lambda K \varphi_\lambda^n + h, \quad \lambda \geq 0. \tag{6.2}$$

As first we assume all the terms in the problem to be nonnegative. We suppose that the following condition is met:

$$q_n = n\lambda\|K\| < 1 \tag{6.3}$$

and for $\lambda = \lambda_0$ there exists such a solution φ_{λ_0} that $\|\varphi_{\lambda_0}\| < 1$. In so doing the solution to equation (6.1) that satisfies the additional condition $\|\varphi_\lambda\| < 1$ is unique for $\lambda \leq \lambda_0$. Indeed, let us assume that there are two such solutions $\varphi_\lambda^{(1)}$ and $\varphi_\lambda^{(2)}$.

Then the following relation is valid:

$$\varphi_\lambda^{(1)} - \varphi_\lambda^{(2)} = \lambda K\left[(\varphi_\lambda^{(1)} - \varphi_\lambda^{(2)})\sum_{k=0}^{n-1}(\varphi_\lambda^{(1)})^k(\varphi_\lambda^{(2)})^{(n-1-k)}\right] = K_\varphi[\varphi_\lambda^{(1)} - \varphi_\lambda^{(2)}].$$

It is obvious that with the constraint (3.4) we have $\|K_\varphi\| < 1$ and therefore $\varphi_\lambda^{(1)} \equiv \varphi_\lambda^{(2)}$.

One can easily show that the solution φ_λ exists and monotonously increases with λ for $\lambda \leq \lambda_0$. Really, if $\lambda_1 \leq \lambda_2 \leq \lambda_0$, then

$$\lambda_1 K\varphi_{\lambda_2}^n + h \leq \varphi_{\lambda_2}.$$

Thus, iterations of equation (6.1) beginning with $\varphi_0 \equiv \varphi_{\lambda_2}$ monotonously decrease and converge to the solution for $\lambda = \lambda_1$ and $\varphi_{\lambda_1} \leq \varphi_\lambda$.

Note that the Monte Carlo algorithms for solving (6.1) were first constructed in a more general form in [16] on the basis of the connection between integral equations with power nonlinearity and branching Markov chains. These chains are constructed in the same manner as those in Section 1.2 (i.e., for $n = 1$), with the difference that after the next transition $x \to x'$, n independent trajectories are constructed from the point x', with the weight previously multiplied by a weight factor, which we assume here to be bounded:

$$q(x,x') = \frac{k(x,x')}{p(x,x')} \le c < +\infty. \tag{6.4}$$

The basic estimate can be defined by the following recurrent formula

$$\xi_0 = h_0 + q(x_0,x_1)\prod_{k=1}^{n}\xi_1^{(k)}, \tag{6.5}$$

where $\{\xi_1^{(k)}\}$ are independent (in the aggregate) values of the estimate that correspond to the above-mentioned trajectories originating at the point x'. According to the above conclusions condition (6.3) provides the termination of a branching trajectory with the probability 1 (and even the finiteness of the average number of branches that is necessary in order to make calculations possible) and thereby provides a possibility to continue recursion (6.5) up to the end of the trajectory. In addition, for small values of λ the value of $\|E\xi\|$ is small as well, since the following relation is valid, provided that the inequality $\lambda n < 1$ is true:

$$\xi_\lambda \le \frac{\|h\|}{1-\lambda n}$$

Theorem 6.1. *If $k(x,x') \ge 0$, $h(x) \ge 0$, $\lambda_0 \ge 1$ and conditions (1.11), (6.3), (6.4) are met, then recursion (6.5) defines an estimate ξ such that*

$$E\xi \equiv \varphi, \quad \|\varphi\| < 1.$$

Proof. As with $n = 1$ the direct averaging of relation (6.5) for $n > 1$ gives equation (6.1). Therefore if the inequality $\|E\xi\| < 1$ is valid, the conclusion of the theorem is true. Let us now suppose that $\|E\xi\| \ge 1$. The properties of the function $\|E\xi(\lambda)\|$ mentioned in Section 1.2 are valid for $n > 1$ as well. Therefore one can find such a $\lambda < 1$ that the following inequality is valid:

$$\|\varphi\| < \|E\xi(\lambda)\| \le 1.$$

In this case $E\xi(\lambda) \equiv \varphi_\lambda$ and $\|\varphi_\lambda\| \le \|\varphi\|$, since $\lambda < 1$. The contradiction obtained proves the theorem. □

The unbiasedness of the obtained estimate is proved in [16] with the more general suggestion of the convergence of iterations for equation (6.1), i.e., the estimate of an iterative solution is considered. In so doing one uses a cumbersome representation of the estimate as a whole and the Neumann series for equation (6.1) (see [17]). Note that the estimate ξ can be realized numerically by using, for instance, the well-known lexicographic scheme (see, for example, [17]).

The direct averaging (as with $n = 1$) of recurrent representation of ξ^2 in (6.5) gives the following equation (see also [48]) for $\psi \equiv E\xi^2$:

$$\psi = h(2\varphi - h) + K_p\psi^n. \tag{6.6}$$

Here, as in Section 1.3 K_p is an integral operator with the kernel $k^2(x,x')/p(x,x')$. For the function $\mathrm{E}\xi^2$ one can obtain a statement similar Theorem 6.1, if one additionally demands that condition (6.3) be met with the substitution $K \to K_p$, $\varphi \to \psi$ and takes equation (6.6) instead of (1.18).

The transition to the case with alternating signs is performed in the following way as with $n = 1$ (see Section 1.3). Initially the reasoning is made for the problem with the kernel $k(x,x')$ and the free term $h(x)$. Thus, one obtains a majorant for ξ and uses it in averaging recursion (6.5). As a result one obtains the function $\mathrm{E}\xi$ that satisfies equation (6.1) and inequality $\|\mathrm{E}\xi\| < 1$, i.e., the solution sought for. Arguments for $\mathrm{E}\xi^2$ are similar.

Now let us consider the minimization of the value of $\mathrm{E}\xi^2$. As in Section 1.3, one can easily find that in order to make the modelling algorithm optimal in this sense of the Bellman principle it is necessary that the following relations be valid:

$$\begin{aligned} p(x,x') &= \frac{|k(x,x')|g^{n/2}(x')}{\int_X |k(x,x')|g^{n/2}(x')dx'} \\ g^{1/2}(x) &= \left\{h(x)[2\varphi(x) - h(x)] + \left[\int_X |k(x,x')|g^{n/2}(x')dx'\right]^2\right\}^{1/2}. \end{aligned} \tag{6.7}$$

The sufficiency of these relations for constructing the optimal variant provided that the iterations of equation (6.7) converge is proved in [5]; in this case $\mathrm{E}\xi^2 \equiv g$. If all the terms in the problem are nonnegative, then one can easily check that for $\lambda_0 > 1$ the operator of the right-hand side of equation (6.7) is an operator of contraction in the ball $\|g^{1/2}\| \le 1$. The unique solution of (6.7) in this ball is the function φ, i.e., $\mathrm{E}\xi^2 = \varphi^2$ and $\mathrm{D}\xi = 0$ (see also [5]).

Note that by using recursion (6.5) the results obtained can be especially easily extended to the equation of a more general form:

$$\varphi = \sum_i K_i\varphi^i + h.$$

6.2. SOLUTION OF DIRICHLET PROBLEM FOR ELLIPTIC EQUATIONS

Let us consider the following boundary-value problem in a bounded domain $D \subset R_3$ with a piecewise smooth boundary Γ:

$$\Delta u + u^n = 0, \quad u\big|_\Gamma = \psi, \quad n \ge 2, \tag{6.8}$$

where $\psi \in C(\Gamma)$. The function u is considered in $C_2(D) \cap C(\bar{D})$ [45]. Along with (6.8) we consider the parametrized problem

$$\Delta u_c + u_c^n = 0, \quad u_c\big|_\Gamma = c\psi, \quad c \ge 0. \tag{6.9}$$

On the strength of the results in [45] one can argue that there exists $c^{(0)} > 0$ such that for $c \le c^{(0)}$ a solution of problem (6.9) is existent and unique in the ball:

$$n\|u_c^{n-1}\| \le c^*, \quad \|u_c^{n-1}\| = \max_{r\in D} |u_c^{n-1}(r)|,$$

where $c^* > 0$ is the first eigenvalue of the problem

$$\Delta u + cu = 0, \quad u\big|_\Gamma = 0.$$

As in Section 6.1, here the uniqueness can be easily established by considering the difference between $u_{c,1}$ and $u_{c,2}$ which are solutions of problem (6.9). If $\psi \ge 0$, then the monotonous increase of u_c and therefore $\|u_c\|$ with c is established as in Section 6.1 by means of the iterative transition from u_{c_2} to u_{c_1} $[c_1 < c_2 < c^{(0)}]$ on the basis of an integral equation [45] that is equivalent to problem (6.9) and determined by the Green function for D.

For sufficiently small values of c the parametric derivatives $u^{(k)}$ are the solutions of the following boundary-value problems:

$$\Delta u' + nu^{n-1}u' = 0, \quad u'\big|_\Gamma = \psi, \quad \Delta u^{(k)} + (u^n)^{(k)} = 0, \quad u^{(k)}\big|_\Gamma = 0, \quad k > 1. \tag{6.10}$$

Some conditions for the validity of relations (6.10) will be established below through the analysis of the corresponding probabilistic representations.

In present section we construct Monte Carlo estimates in 'random walk on spheres and balls' (see Section 3.4) for the problems involved. Branching occurs when the trajectory hits a ball. Branching chains of this kind with additional absorption were considered in [18] in order to estimate the solution of the equation $\Delta u = u^2$. New estimates presented here are characterized by pure multiplication. They call for special investigation, using the result of Section 3.4.

Let us introduce the notation: $d(r)$ is a distance between the point r and the boundary Γ, Γ_ε is the ε-vicinity of the boundary, $S(r)$ is the sphere of radius $d(r)$ with the center at the point r, $D(r)$ is the corresponding ball.

By analogy with Section 3.4 we consider the following integral representation of the solution:

$$u(r) = \frac{1}{4\pi d^2(r)} \int_{S(r)} u(r(s))ds + \int_{D(r)} u^n(r')G_r(r')dr', \quad r \in D\backslash\Gamma_\varepsilon. \tag{6.11}$$

For $r \in \Gamma_\varepsilon$ one uses the identical representation $u(r) \equiv u(r)$. Its approximation by the boundary condition ($u \approx \tilde{u}$) leads to the ε-biased estimate (see Section 3.4). Here $G_r(r')$ is the Green's function for a ball with the pole at the centre of the ball:

$$G_r(r') = \frac{1}{4\pi}\left[\frac{1}{|r-r'|} - \frac{1}{d(r)}\right], \quad |r-r'| \le d(r).$$

In order to construct the Monte Carlo estimate for problem (6.8), by analogy with Section 3.4 one realizes random walk on spheres and balls from the point $r_0 = r$. In so doing a new point r_{i+1} is chosen uniformly on the sphere $S(r_i)$ with the probability $1 - c_0 d^2(r_i)/6$, and the weight Q is multiplied by $q(r_i, c_0) = [1 - c_0 d^2(r_i)/6]^{-1}$; the point r_{i+1} is chosen within the ball $D(r_i)$ with the complementary probability $c_0 d^2(r_i)/6$ according to the density $6d^{-2}(r_i)G_{r_i}(r')$ and the weight is multiplied by c_0^{-1}. In the last case, n-multiple branching is performed, i.e., n conditionally-independent random walk trajectories are constructed from the point r_i. Here $c_0 < 0.6079c^*$ (see Section 3.4). When hitting Γ_ε, the corresponding branch of a trajectory terminates and the quantity Qu (or $Q\tilde{u}$ if one goes to the real ε-biased estimate) is added to the value of the estimate. Further (see Theorem 6.3) we will show that under certain conditions the branching chain involved terminates with the probability 1. The constructed estimate, is of the following recurrent form:

$$\xi_0 = \begin{cases} [1 - c_0 d^2(r)/6]^{-1}\xi_1, & P = 1 - c_0 d^2(r)/6 \\ \left(\prod_{k=1}^{n} \xi_1^{(k)}\right)/c_0, & P = c_0 d^2(r)/6. \end{cases}$$

Continuing the recursion up to the end of the branching trajectory, we find the following probabilistic representation of the Neumann series for problem (6.8) (such constructions are fully considered in details, for example, in [18]). The next theorem assumes the branching chain to terminate with the probability 1 and $\varepsilon = 0$.

Theorem 6.2. *If* $c^{(0)} \geq 1$ *and* $c_0 < 0.6079c^*$ *then* $\mathrm{E}\xi \equiv u$, *with* $n\|u^{n-1}\| < c^*$.

Proof. At first we assume that $\psi \geq 0$. Then any rearrangements of summations and preliminary conditional averaging procedures are allowed in the process of averaging, since all the elements of the estimate ξ are nonnegative. Thus, $\mathrm{E}\xi = \mathrm{E}\zeta$ and

$$\zeta_0 = \begin{cases} [1 - c_0 d^2(r)/6]^{-1}\zeta_1, & P = 1 - c_0 d^2(r)/6 \\ (\mathrm{E}\xi_1)^{n-1}\zeta_1/c_0, & P = c_0 d^2(r)/6. \end{cases}$$

Therefore, if the inequality

$$\|(\mathrm{E}\xi)^{n-1}\| < c^*/n$$

is valid, then taking into account the inequality $c_0 < 0.6079c^*$, one can state on the basis of Section 3.4 that the function $\mathrm{E}\xi$ is the solution of problem (6.8). Let us now assume

$$\|(\mathrm{E}\xi)^{n-1}\| \geq c^*/n. \tag{6.12}$$

As in section 1.2, here we arrive at a monotonous and continuous (with respect to c) increase of the function $\|\mathrm{E}\xi(c)\|$ up to $+\infty$, where $\xi(c)$ is the estimate of the solution of the parametrized problem. Hence if inequality (6.12) and the inequality $c^{(0)} > 1$ one can find such $c < 1$ that the following relation holds:

$$\|u^{n-1}\| < \|(\mathrm{E}\xi(c))^{n-1}\| < c^*/n.$$

However, in the case $\mathrm{E}\xi(c) \equiv u_c$ and $\|u_c^{n-1}\| \leq \|u^{n-1}\|$, since $c < 1$. The contradiction obtained proves the theorem for $\psi \geq 0$. Going to the case with alternating signs is performed in the same manner as in Section 1.2, by using the estimate ξ for problem (6.9) as the majorant with the substitution $\psi \to |\psi|$. Theorem 6.2 is proved. □

Below we give a sufficient condition for the finiteness of the average number of branches that is necessary in order to make calculations possible.

Theorem 6.3. *If* $c_0 < 0.6079c^*$ *and* $n < c^*$ *then the average number of branches that have reached the boundary in the random walk under consideration for* $\varepsilon = 0$ *coincides with the solution of the problem* $\Delta u + nu = 0$, $u\big|_\Gamma = 1$.

Proof follows immediately from Section 3.4 with allowance made for the fact that branching is, on average, equivalent to multiplying the auxiliary weight by n. □

Theorem 6.3 gives a logarithmic estimate of the complexity: $S_\varepsilon \asymp |\ln \varepsilon|$, with the known suppositions of the regularity of the boundary Γ [18,39]. Note that with the uniform boundedness of the derivatives of the solution with respect to coordinates in D, in case $r \in \Gamma_\varepsilon$ the substitution of $u(r)$ for the value of ψ at the point of the boundary nearest to r leads to the ε-biased estimate (see Section 3.4). In view of Theorem 6.3 we will further assume that $c_0 > 2$.

Let us now consider the quantity $\mathrm{E}\xi^2$. In this connection we need to consider Monte Carlo estimates for the problem

$$\Delta u_1 + u_1\Big(u_1^{n-1} + \frac{1}{2}c^*\Big) = 0, \quad u_1\big|_\Gamma = \psi. \tag{6.13}$$

By analogy with problem (6.8), on the basis of the results in [45] we assume that for $c \le c_1^{(0)}$ the solution $u_{1,c}$ of the parametrized problem (6.13) is existent and unique in the ball

$$2n\|u_{1,c}^{n-1}\| < c^*$$

and also monotonously increases with c.

We also need to use the inequality

$$\frac{1}{1 - c_0 d^2(r)/6} < \frac{d\sqrt{c^*/2}}{\sin(d\sqrt{c^*/2})} = s(d, c^*/2) \tag{6.14}$$

which is valid for $d\sqrt{c^*} \le \pi$, provided that $c_0/c^* \le 0.39$. This inequality is proved by considering the difference of quantities inverse to the sides of (6.14), divided by $x^2 = c^* d^2$; the derivative of the obtained series in powers of x^2 is negative and the series itself vanishes at $c_0/c^* = 0.39...$, $x = \pi$. In addition, we need the inequality

$$\frac{s(d, c^*/2)}{s((d-t), c^*/2)} < 1.56 s(t, c^*) < c_0 s(t, c^*) \tag{6.15}$$

that is valid, provided that $c_0 > 2$ for $0 \le t \le d$ at $d\sqrt{c^*} < \pi$. On substitution of $x = d\sqrt{c^*/2}$ inequality (6.15) takes the form:

$$t(x) = \frac{x}{\sin(x)} \cdot \frac{\sin(x-y)}{x-y} < 1.56, \quad 0 < y \le x \le \pi/\sqrt{2}.$$

By calculating the logarithmic derivative one can easily show that the function $t(x)$ increases monotonously with x.

On the other hand, direct calculations show that

$$\frac{\pi\sqrt{2}}{\sin(\pi/\sqrt{2})} \cdot \frac{\sin(\pi/\sqrt{2} - y)}{\pi/\sqrt{2} - y} < 1.56, \quad y < \pi/\sqrt{2}.$$

Therefore inequality (6.15) does hold.

Theorem 6.4. *If* $c_1^{(0)} \ge 1$ *and* $c_0 < 0.39c^*$, *then* $\|\mathrm{E}\xi^2\| < \|u_1\|$, *with* $\|u_1^{n-1}\| < c^*/(2n)$.

Proof. The following representation is valid for ξ^2:

$$\xi_0 = \begin{cases} [1 - c_0 d^2(r_0)/6]^{-2}\xi_1, & P = 1 - c_0 d^2(r_0)/6 \\ \left[\prod_{k=1}^{n} (\xi_1^{(k)})^2\right] / c_0^2, & P = c_0 d_0^2(r_0)/6. \end{cases}$$

In view of inequality (6.14), the value of ξ^2 is majorized by the quantity

$$\zeta_0^2 = \begin{cases} s(d_0, c^*/2)[1 - c_0 d_0^2/6]^{-1}\zeta_1^2, & P = 1 - c_0 d_0^2/6 \\ \dfrac{s(d_0, c^*/2)}{c_0 s((d_0 - t), c^*/2)} \left[\prod_{k=1}^{n} (\zeta_1^{(k)})^2\right] / c_0, & P = c_0 d_0^2/6. \end{cases}$$

As in the proof of Theorem 6.2 one can here replace $\prod_{k=1}^{n-1}(\zeta_1^{(k)})^2$ by $(E\zeta_1^2)^{n-1}$. It is easy to verify that the resulting expression determines the estimate of the solution of problem

$$\Delta u + \frac{1}{2}c^* u = -u(E\zeta^2)^{n-1}/c_0, \quad u\big| = \psi^2$$

by the method described in Section 3.4 and in view of (6.14), (6.15) the corresponding expectation is finite, if $(E\zeta^2)^{n-1} < c_0 < 0.39c^*$. Then one repeats the proof of Theorem 6.2 almost word for word. □

The estimates of the parametric derivatives $u^{(k)}$ for $\psi \geq 0$ can be obtained by the direct differentiation of the estimate ξ (going to the case with alternating signs can be made as before by using majorant derivatives). As this takes place, one obtains randomized estimates of the solutions to Helmholtz equations (6.10). On this basis one can obtain sufficient conditions for finiteness of such derivatives. For example, for u' we have the following probabilistic representation:

$$\xi' = \begin{cases} [1 - c_0 d^2(r)/6]^{-1}\xi', & P = 1 - c_0 d^2(r)/6 \\ n\xi_1' \left(\prod_{k=1}^{n-1} \xi_1^{(k)}\right) / c_0, & P = c_0 d^2(r)/6 \end{cases}$$

and the value of $\|E\xi'\|$ is obviously finite, if $n\|u^{n-1}\| < c_0 < 0.6079c^*$. We also find from Section 3.4 that $\|E\xi'^2\| < +\infty$ if

$$n\|(E\xi^2)^{(n-1)/2}\| < c_0 < 0.488c^*.$$

Note that for $c = 0$ from (6.10) there follows the sequence of Poisson's equations that can be solved using standard randomized estimates with a finite variance.

The results obtained can be extended to equations of the form:

$$\Delta u_1 + u_1 \sum_{k=0}^{n} a_k u^k = 0, \quad u\big|_\Gamma = \psi > 0,$$

and to the similar systems of equations, for instance, to the system:

$$\Delta u_1 + a_1 u_1 u_2 = 0, \quad \Delta u_2 + a_2 u_1 u_2 = 0.$$

To extend these results to nonhomogeneous equations, for instance, to the equation $\Delta u + u^n = -g$, it is necessary to use the parametrization of the type:

$$\Delta u_c + u_c^n = -cg, \quad u_c\big|_\Gamma = c\psi.$$

Here only the method of investigating the variance changes essentially.

The results obtained can be interpreted in the following way, without regard to the properties of the solution of parametrized problem (6.8) which were formulated at the beginning of the paper. The continuous (in c) function $u_c = \mathrm{E}\xi$ (for $\varepsilon = 0$ and $\tilde{u} \equiv c\psi$), which is the solution of problem (6.8)with the substitution $\psi \to c\psi$, is found, provided that $\|u_c^{n-1}\| < c^*$ and $n < 0.6079c^*$. The equality $u \equiv 0$ is valid for this function at $c = 0$. The parametric derivative u^k is the solution of the corresponding Dirichlet problem (6.8) under the additional constraints of the form $n\|u^{n-1}\| < c^*$ for $k = 1$.

6.3. MINIMIZATION OF COST OF MONTE CARLO METHODS IN ITERATIVE SOLUTION OF NONLINEAR PROBLEMS

Let us assume that the solution $f^*(x)$ to the nonlinear operator (integral, for example), equation

$$f = C(f)$$

can be found by the iterative method starting with the approximation $f_0(x)$; in this case, we have

$$\| Gf_n - f^* \|_{L_2} \leq q \| f_n - f^* \|_{L_2}, \quad q < 1. \tag{6.16}$$

Assume that Eq. (6.16) is also valid for approximations $f_n^{(\varepsilon)}$ of the frequency polygon type estimated by the Monte Carlo method. The histogram method is considered here because it is most widely used in the iterative solution of nonlinear transport equations by the Monte Carlo method.

By $\tilde{f}_n^{(\varepsilon)}$ denote the statistical estimate of the frequency polygon and by E_{n-1} denote the conditional mathematic expectation under the condition that the value of the $(n-1)$th iteration is fixed and coincides with $\tilde{f}_{n-1}^{(\varepsilon)}$. It means in particular that the following relation is valid:

$$E_{n-1}\tilde{f}_n^{(\varepsilon)} = [G\tilde{f}_{n-1}^{(\varepsilon)}]^{(\varepsilon)}.$$

Hence, we can make use of the estimate (see Section 1.5).

$$\begin{aligned} \mathrm{E}\|\tilde{f}_n^{(\varepsilon)} - f^*\|^2 &= \left\|\mathrm{E}\left[(\mathrm{E}_{n-1}\tilde{f}_n^{(\varepsilon)} - f^*)^2 + \frac{\mathrm{D}_{n-1}\tilde{f}_n^{(\varepsilon)}}{N_n}\right]\right\|_{L_2}^2 \\ &\leq ch_n^2 + q^2\mathrm{E}\|\tilde{f}_{n-1}^{(\varepsilon)} - f^*\|_{L_2}^2 + \frac{d}{N_n h_n^k}. \end{aligned} \tag{6.17}$$

Here, k is the number of measurements of the space X, h_n is the grid step size used at the nth iteration, and N_n is the corresponding size of the sample in the Monte Carlo method. Then successively manipulating with Eq. (6.17), by induction we obtain

$$E(\tilde{f}_n^{(\epsilon)} - f^*)^2 \leq \sum_{i=1}^{n} \left(ch_i^2 + \frac{d}{h_i^k N_i} \right) q^{2(n-i)} + q^{2n} \parallel f_0 - f^* \parallel^2 . \tag{6.18}$$

If we choose $n \approx c \ln \delta / \ln q$ for $c > 1$, we can obviously ignore the last term in Eq. (6.18). After that the cost minimization problem takes the form

$$S = t \sum_{i=1}^{n} N_i \rightarrow \min_{\{N_i, h_i\}}, \quad \sum_{i=1}^{n} \left(ch_i^2 + \frac{d}{h_i^k N_i} \right) q^{2(n-i)} = \delta^2. \tag{6.19}$$

Substituting N_n from the last equality we obtain the following expression of the value minimized:

$$\sum_{i=1}^{n-1} N_i + \frac{d}{h_n^k \left[\delta^2 - \sum_{i=1}^{n-1} \left(ch_i^2 + \frac{d}{h_i^k N_i} \right) q^{2(n-i)} - ch_n^2 \right]}. \tag{6.20}$$

By equating the derivatives of expression of Eq. (6.20) to zero over the values h_i we obtain the relations

$$2ch_i - \frac{kd}{N_i h_i^{-(k+1)}} = 0, \quad i = 1, 2, ..., n-1.$$

Hence,

$$N_i = \frac{kd}{2c} h_i^{-(k+2)}, \quad i = 1, 2, ..., n-1. \tag{6.21}$$

From Eq. (6.21) we have

$$ch_i^2 + \frac{d}{h_i^k N_i} = \frac{k+2}{k} ch_i^2.$$

Substituting this expression and Eq. (6.21) into Eq. (6.20) we obtain the expression

$$\frac{d}{c} \left[\sum_{i=1}^{n-1} \frac{k}{2h_i^{k+2}} + \frac{1}{h_n^k \left(\frac{\delta^2}{c} - \frac{k+2}{k} \sum_{i=1}^{n-1} h_i^2 q^{2(n-i)} - h_n^2 \right)} \right].$$

Equating the derivatives of the last expression to zero over h_i, $i = 1, ..., n-1$ and h_n we obtain correspondingly the equalities

$$k^2 h_n^k \left(\frac{\delta^2}{c} - \frac{k+2}{k} \sum_{i=1}^{n-1} h_i^2 q^{2(n-i)} - h_n^2 \right) = 4 h_i^{k+4} q^{2(n-i)}$$

$$\frac{\delta^2}{c} - \frac{k+2}{k} \sum_{i=1}^{n-1} h_i^2 q^{2(n-i)} - h_n^2 = \frac{2}{k} h_n^2. \tag{6.22}$$

Substituting the second equality into the first one we obtain

$$h_i = h_n q^{-2(n-i)/(k+4)}, \quad i = 1, 2, ..., n-1. \tag{6.23}$$

Hence,

$$ch_i^2 + \frac{d}{h_i^k N_i} = \frac{k+2}{k} c q^{-4(n-i)/(k+4)} h_n^2, \quad i = 1, \ldots, n-1.$$

Substituting expressions of Eq. (6.23) into Eq. (6.22) we have

$$h_n = \left[\frac{k}{c(k+2)(1+Q)}\right]^{1/2} \delta, \quad Q = \sum_{i=1}^{n-1} q^{2(n-i)(k+2)/(k+4)}.$$

Making use of the equality from Eq. (6.19) we then obtain

$$N_n = \frac{d(k+2)^{1+k/2}(1+Q)^{1+k/2}(c/k)^{k/2}}{2\delta^{k/2}}. \tag{6.24}$$

In addition, Eq. (6.21), Eq. (6.23) and Eq. (6.24) imply that

$$N_i = N_n q^{2(n-i)(k+2)/(k+4)}.$$

It is easy now to obtain the minimal value of the cost:

$$S^* = N_n(1+Q) = \frac{d(k+2)^{1+k/2}(c/k)^{k/2}(1+Q)^{2+k/2}}{2\delta^{k+2}} \lesssim \frac{d(k+2)^{1+k/2}(c/k)^{k/2}}{2(1-q')^{2+k/2}\delta^{k+2}},$$

where

$$q' = q^{2(k+2)/(k+4)}.$$

It is interesting to note that the order of the minimal cost over δ is the same here as in the solution of the linear integral equation by the histogram method.

Note that the results obtained can be easily generalized to the case of prolongation of order $2m$ and the cost of one sample of the form t/h^r. In this case, we have the following problem:

$$t \sum_{i=1}^{n} \frac{N_i}{h_i^r} \to \min, \quad \sum_{i=1}^{n} \left(ch_i^{2m} + \frac{d}{h_i^k N_i}\right) q^{2(n-k)} = \delta^2.$$

Making the replacement

$$N_i' = N_i/h_i^r, \quad h_i' = h_i^m$$

we arrive at the problem considered with the replacement $k \to (r+k)/m$. For example, if $r = 0$ and $m = 2$, the order of the minimal cost is defined by the quantity $\delta^{-2-k/2}$ as in the case of using the frequency polygon for solving the linear integral equation. We can note that the role of an increase in the order of prolongation grows, with the order of growth of the variance and the cost of one statistical sample increased.

Now consider the cost minimization problem when solving a nonlinear transfer equation in total by analogy with Section 5.7. Assume that coefficients of equation (5.16) are dependent on φ (see Section 5.7) continuously with respect to the metric C, i.e., $\varphi = G(\varphi)$ and, in addition, the solution of the latter functional equation can be found by the iterative method, starting with the approximation φ_0; in this case we have

$$\|G\varphi_n - \varphi^*\|_{L_2(H)} \leq q\|\varphi_n - \varphi\|_{L_2(H)}, \quad q < 1. \tag{6.25}$$

Assume that (6.25) is also valid for approximations $\varphi_n^{(h)}$ of the frequency polygon type estimated by the free-path Monte Carlo method.

Here

$$\mathrm{E}_{n-1}\tilde{\varphi}_n^{(h)} = [G\tilde{\varphi}_{n-1}^{(h)}]^{(h)}$$

Hence

$$\begin{aligned} \mathrm{E}\|\tilde{\varphi}_n^{(h)} - \varphi^*\|^2 &= \mathrm{E}\|\varphi^* - \mathrm{E}_{n-1}\tilde{\varphi}_{n-1}^{(h)}\|^2 + \frac{\mathrm{D}_{n-1}\tilde{\varphi}_n^{(h)}}{N_n} \\ &\leq \frac{d}{N_n} + ch_n^2 + q^2\mathrm{E}\|\tilde{\varphi}_{n-1}^{(h)} - \varphi^*\|^2. \end{aligned} \tag{6.26}$$

Then successively manipulating with (6.26), by induction we obtain

$$\mathrm{E}\|\tilde{\varphi}_n^{(h)} - \varphi^*\|^2 \leq \sum_{i=1}^{n} \Big(\frac{d}{N_i} + ch_i^2\Big) q^{2(n-i)} + q^{2n}\mathrm{E}\|\varphi_0 - \varphi^*\|^2.$$

If we choose $n \asymp \ln\delta/\ln q$, we can obviously ignore here the last term. After that the cost minimization problem takes the form

$$S = t\sum_{i=1}^{n} \frac{N_i}{h_i} \to \min_{\{N_i,h_i\}}, \quad \sum_{i=1}^{n} \Big(\frac{d}{N_i} + ch_i^2\Big) q^{2(n-i)} = \delta^2. \tag{6.27}$$

As previously for (6.19), it is possible to obtain the solution of problem (6.27):

$$h_n^* = \delta[3cQ]^{-1/2}; \quad 2N_n^* = 3^{3/2}dc^{1/2}Q^{3/2}\delta^{-3}; \quad h_i^* = h_n^* q^{\frac{-2(n-i)}{5}};$$

$$N_i^* = N_n^* q^{\frac{6(n-i)}{5}}; \quad Q = 1 + \sum_{i=1}^{n-1} q^{\frac{6(n-i)}{5}}.$$

The corresponding value of the cost is $S^* \asymp \delta^{-3}$.

By the use of the imbedding theorem it is possible to obtain the similar results with respect to the metric C, when solving the nonlinear equation (5.16). As in Section 5.7 we have

$$\begin{aligned} \mathrm{E}\|\tilde{\varphi}_n^{(h)} - \varphi^*\|_C^2 &\leq \mathrm{E}\Big(\int_H (\tilde{\varphi}_n^{(h)} - \varphi^*)^2(z)dz + \int_H \big((\tilde{\varphi}_n^{(h)})' - (\varphi^*)'\big)^2(z)dz\Big) \\ &\leq \frac{d_1}{N_n h_n} + c_1 h_n + \mathrm{E}\int_H \{(G(\tilde{\varphi}_n^{(h)}) - \varphi^*)^2(z) + ((G(\tilde{\varphi}_n^{(h)}))' - (\varphi^*)')^2(z)\}dz. \end{aligned}$$

By virtue of the transfer equation the derivative is expressed in the form of a linear functional of the solution and can be estimated on the basis of the perturbation theory (see, for instance, [27]). Therefore

$$\mathrm{E}\|\tilde{\varphi}_n^{(h)} - \varphi^*\|_C^2 \leq \frac{d_1}{N_n h_n} + c_1 h_n + q_1^2\mathrm{E}\|\tilde{\varphi}_{n-1}^{(h)} - \varphi^*\|_{L_2(H)}^2, \quad q_1 > q.$$

Further, by analogy with (6.27) we obtain the following cost minimization problem

$$S_1 = t\sum_{i=1}^{n} \frac{N_i}{h_i} \to \min_{N_i,h_i}; \quad \sum_{i=1}^{n-1} \Big(\frac{d}{N_i} + ch_i^2\Big) q^{2(n-i-1)} q_1^2 + \frac{d_1}{N_n h_n} + c_1 h_n = \delta^2.$$

Similarly to (6.19) it is possible to obtain here the optimal relations:

$$\frac{3c_1}{2}h_n^* + 3c\frac{dc_1^2}{d_1c^2 16}(h_n^*)^{8/5}Q = \delta^2; \quad h_i^* = (h_n^*)^{4/5}\left(\frac{dc_1^2}{d_1c^2 16}\right)^{1/5} q_1^{-2/5} q^{\frac{-2(n-i)}{5}};$$
$$N_i^* = \left[\frac{N_n^* c_1 q_1}{2d_1}\right]^{4/5} q^{\frac{4(n-i-1)}{5}} \frac{d}{2c}\left[\frac{d_1c^2 16}{dc_1^2}\right]^{2/5}; \tag{6.28}$$
$$Q = q_1^{6/5}\sum_{i=1}^{n-1} q^{\frac{6(n-i-1)}{5}}, \quad N_n^* = \frac{2d_1}{c_1(h_n^*)^2}.$$

The quantity S_1^* asymptotically has the order δ^{-6}, but only the last item in the expression of S_1^* has this order, while the other items have the order $\delta^{-24/5}$.

When using the collision estimate the similar optimal value of S has the order δ^{-8}.

6.4. ITERATIVE SOLUTION OF A MODEL KINETIC EQUATION

We consider the possibilities of applying the above-listed methods for solving a model kinetic equation in the classical problem on heat transfer between two parallel plates. In the case of spatially one-dimensional problem the kinetic equation is written in the form [22]:

$$w_x \frac{\partial f(x,w)}{\partial x} = \nu(x,\mu_\nu)[f_0(x,w)], \tag{6.29}$$

where f_0 is the locally Maxwell distribution function, $\nu(x,\mu_\nu)$ is the collision frequency. In the case of Maxwell molecules we have $\nu(x,\mu_\nu) = An(x)$, where A is a constant. The gas macroparameters that enter into the definitions of f_0 and ν are moments of the distribution function $f(x,w)$:

$$n(x) = \int_V f(x,w)\mathrm{d}w, \quad T(x) = \frac{\int_V w^2 f(x,w)\mathrm{d}w}{3Rn(x)}$$

and represent the density and the temperature of the gas, respectively, R is the gas constant.

The physical formulation of the problem is as follows.

The gas with the average density n_0 is confined between two rigid infinite plates, which coincide with the planes $x = 0$ and $x = d$. The temperatures of the lower plate T_+ and the upper plate T_- are constants. One has to determine the state of gas between the plates for given T_+ and T_- and given degree of gas rarefaction. For the sake of simplicity the law of reflection from the plates is taken to be diffuse with a complete accommodation of energy and momentum [22].

Equation (6.20) is brought to the integral form and the simple iteration method is used for its solution. At each iteration one solves the following integral equation (in dimensionless units):

$$\begin{aligned} f_{m+1}(x,w) = & \int_0^1 \int_V f_{m+1}(x',w') \frac{1}{\pi^{3/2} T_m^{3/2}(x')} \exp\Big(-\frac{w^2}{T_m(x')}\Big) \frac{n_m(x')}{\sqrt{2}Knw} \\ & \times \exp\Big(-\int_0^{|x-x'|} \frac{n_m(x - s\omega_x)}{\sqrt{2}Knw_x} \mathrm{d}s\Big) \delta\Big(\frac{x-x'}{|x-x'|} - \omega_x\Big) \mathrm{d}w'\mathrm{d}x' \\ & + \Big[\frac{1}{\pi^{3/2}} \exp(-w^2)\delta(x') + \frac{1}{\pi^{3/2}T_-^{3/2}} \exp\Big(-\frac{w^2}{T_-}\Big)\delta(x'-1)\Big] \\ & \times \Big(-\int_0^{|x-x'|} \frac{n_m(x - s\omega_x')}{\sqrt{2}Knw_x} \mathrm{d}s\Big) \delta\Big(\frac{x-x'}{|x-x'|} - \omega_x'\Big), \end{aligned} \tag{6.30}$$

where Kn is the Knudsen number, $w^2 = w_x^2 + w_y^2 + w_z^2$, $\omega_x = w_x/|w_x|$. In order to solve the equation one models the particle paths according to the equation kernel and the right-hand side [17]. The path originates at $x = 0, w_x > 0$. The particle return to the plane $x = 0$ is taken as the transition to the absorbing state. The density and temperature are estimated by the histogram method. The space is subdivided into cells by the planes $x_m = m/k$, $m = 0, 1, ..., k$. We use the time estimate which is a special case of the path estimate. The functions $n(x) = 1$ and $T(x) = T_-$ are taken as the initial estimate in the numerical experiments. The iteration process terminates if for two successive approximations the inequality

$$\max_m \frac{|M_m^{n+1} - M_m^n|}{M_m^{n+1}} \le \varepsilon, \quad m = 0, 1, ..., k$$

is valid, where $\varepsilon = 0.09$. More detailed description of the modelling process is given in [22].

It is proved in [22] that the norm of the integral operator in (6.30) is less than unity. This fact provides the solvability of equation (6.30) by the Monte Carlo method. The equations significantly affects following methods are solved using the weighted Monte Carlo methods with the path considered above [44].

Let us now consider the calculation of the derivatives of the solution to the equation under consideration with respect to a parameter [44]. From the calculation data it is obvious that the Knudsen number significantly affects the computational cost of solving the equation. The computation cost increases with decreasing the value Kn. Therefore it is natural to consider the possibility of using the parameter extension method in order to find the solution for small Knudsen numbers. Let us differentiate (6.30) with respect to the Knudsen number. Denote

$$\frac{\partial f}{\partial Kn} = f^{(1)}, \quad \frac{\partial^2 f}{\partial Kn^2} = f^{(2)}.$$

We denote the kernel of equation (6.30) by K^m and the right-hand side by F^m

$$
\begin{aligned}
f^{(1)}_{m+1}(x,w) = & \int_0^1\int_V f^{(1)}_{m+1}(x',w')K^m(x,x',w)\mathrm{d}w'\mathrm{d}x' \\
& + \int_0^1\int_V f_{m+1}(x',w')K^m(x,x',w) \\
& \times\Big[\frac{n^{(1)}_m(x')}{n_m(x')} - Kn^{-1} - \int_0^{|x-x'|}\Big(-\frac{n_m(x-sw_x)}{\sqrt{2}Kn^2w_x} + \frac{n^{(1)}_m(x-sw_x)}{\sqrt{2}Knw_x}\Big)\mathrm{d}s \\
& -\frac{3}{2}\frac{T^{(1)}_m(x')}{T_m(x')} + \frac{w^2T^{(1)}_m(x')}{T^2_m(x')}\Big]\mathrm{d}w'\mathrm{d}x' \\
& + F^m(x,x'w)\Big[-\int_0^{|x-x'|}\Big(\frac{n_m(x-sw_x)}{\sqrt{2}Kn^2w_x} + \frac{n^{(1)}_m(x-sw_x)}{\sqrt{2}Knw_x}\Big)\mathrm{d}s\Big]
\end{aligned}
$$

$$
\begin{aligned}
f^{(2)}_{m+1}(x,w) = & \int_0^1\int_V f^{(2)}_{m+1}(x',w')K^m(x,x',w)\mathrm{d}w'\mathrm{d}x' \\
& +2\int_0^1\int_V f^{(1)}_{m+1}(x',w')K^m(x,x',w) \\
& \times\Big[\frac{n^{(1)}_m(x')}{n_m(x')} - Kn^{-1} - \int_0^{|x-x'|}\Big(-\frac{n_m(x-sw_x)}{\sqrt{2}Kn^2w_x} + \frac{n^{(1)}_m(x-sw_x)}{\sqrt{2}Knw_x}\Big)\mathrm{d}s \\
& -\frac{3}{2}\frac{T^{(1)}_m(x')}{T_m(x')} + \frac{w^2T^{(1)}_m(x')}{T^2_m(x')}\Big]\mathrm{d}w'\mathrm{d}x' + \int_0^1\int_V f_{m+1}(x',w')K^m(x,x',w) \\
& \times\Big\{\Big[\frac{n^{(1)}_m(x')}{n_m(x')} - Kn^{-1} - \int_0^{|x-x'|}\Big(-\frac{n_m(x-sw_x)}{\sqrt{2}Kn^2w_x} + \frac{n^{(1)}_m(x-sw_x)}{\sqrt{2}Knw_x}\Big)\mathrm{d}s \\
& -\frac{3}{2}\frac{T^{(1)}_m(x')}{T_m(x')} + \frac{w^2T^{(1)}_m(x')}{T^2_m(x')}\Big]^2 + \frac{n^{(2)}_m(x')}{n_m(x')} - \frac{(n^{(1)}(x'))^2}{(n_m(x'))^2} + Kn^{-2} \\
& -\int_0^{|x-x'|}\Big(2\frac{n_m(x-sw_x)}{\sqrt{2}Kn^3w_x} - 2\frac{n^{(1)}_m(x-sw_x)}{\sqrt{2}Kn^2w_x} + \frac{n^{(2)}_m(x-sw_x)}{\sqrt{2}Knw_x}\Big)\mathrm{d}s \\
& -\frac{3}{2}\frac{T^{(2)}_m(x')}{T_m(x')} + \frac{3}{2}\Big(\frac{T^{(2)}_m(x')}{T_m(x')}\Big)^2 + \frac{w^2T^{(2)}_m(x')}{T^2_m(x')} - 2\frac{w^2(T^{(2)}_m)^2(x')}{T^3_m(x')}\Big\}\mathrm{d}w'\mathrm{d}x' \\
& + F^m(x,x'w)\Big\{\Big[-\int_0^{|x-x'|}\Big(\frac{n_m(x-sw_x)}{\sqrt{2}Kn^2w_x} + \frac{n^{(1)}_m(x-sw_x)}{\sqrt{2}Knw_x}\Big)\mathrm{d}s\Big]^2 \\
& -\int_0^{|x-x'|}\Big(2\frac{n_m(x-sw_x)}{\sqrt{2}Kn^3w_x} - 2\frac{n^{(1)}_m(x-sw_x)}{\sqrt{2}Kn^2w_x} + \frac{n^{(2)}_m(x-sw_x)}{\sqrt{2}Knw_x}\Big)\mathrm{d}s\Big\}.
\end{aligned}
$$

Let us consider the possibility of using these relations for calculating the derivatives by the Monte Carlo method. It follows from the results in [22] that $\|K^m\| < 1$. From the statements of the problem it follows that the integral operators with the kernels of the last type are bounded. In addition, for sufficiently small $\varepsilon > 0$ and all the values of Kn from the interval $Kn_0 - \varepsilon < Kn' < Kn_0 + \varepsilon$ the following inequality is valid:

$$|k^m(x',x,Kn)| \leq (1 + c_0\varepsilon)k_\varepsilon(x,x'),$$

where $k_\varepsilon(x,x')$ is the kernel of the equation that corresponds to the Knudsen number $Kn_\varepsilon = Kn_0 + \varepsilon$. Thus, by virtue of Section 2.2 one can use the above triangular system for calculating the derivatives.

Section 5.3 shows that for the free-path estimate the variance of estimates of the components of the solution to the system is finite if the inequality $\|K_{p,ii}\| < 1$ is valid. In our case a direct modelling is realized and the condition is met. Therefore, the variance is finite.

We also consider the possibilities of using the perturbation theory algorithms [27]. It is natural to consider the Knudsen number as a parameter in the method of the existing parameter. However, when constructing the system of equations one finds out that the norm of their integral operators exceeds unity. When using the method of an introduced parameter one obtains a system of integral equations with the operator norm less than unity. But in this model additional restrictions on the initial estimate emerge, that worsen the algorithm convergence. Since the right-hand side becomes more complicated with each step and, correspondingly, the computational cost of solving the equation increases, this method proves to be very laborious for this equation and we do not study this method in detail.

In the course of numerical experiments for the model kinetic equation we study the domain of numbers $Kn \sim 1$. We take the functions $n(x) = 1$, $T(x) = T_-$ as the initial estimate in the numerical experiments, and when calculating the derivatives with respect to the Knudsen numbers we use the functions $n^{(1)}(x) = 0$, $n^{(2)}(x) = 0$, $T^{(1)}(x) = 0$, $T^{(2)}(x) = 0$. In order to decrease the computational cost we use optimal relations (6.28). The constants used in these formulae are estimated on the basis of preliminary calculations. Table 6.1 lists the density and temperature estimates for the parameters $Kn = 0.5$ and $T_- = 0.25$. The following notation is used: $n_{\text{opt}}(x)$, $T_{\text{opt}}(x)$, $n(x)$, $T(x)$ are the density and temperature calculated with and without use of the optimal relations, respectively; σ is the mean square error estimate, δ_n and δ_T are relative errors of the form:

$$\max_x \left| \frac{n_{\text{opt}}(x) - n(x)}{n(x)} \right|.$$

The number of iterations in various regimes changes form 2 to 9 and increases with decreasing of the Knudsen number and temperature of the upper plate. Using the optimal relations allows one to reduce the computing time by several times in comparison with [22]. In particular, by 6 times for the equation with the parameters $T_- = 0.25$ and $Kn = 0.5$, and by 5 times for $T_- = 0.25$ and $Kn = 0.2$.

Using the parameter expansion method we calculate the derivatives with respect to the Knudsen number in the domain $Kn \sim 1$ for $T_- = 0.25$. For example, we calculate the first and second derivatives for $Kn = 1$. Table 6.2 lists the calculation data.

Table 6.1. Results of solving problem (6.29) by the simple iteration method

x	$n \pm \sigma$	$T \pm \sigma$	$n_{\text{opt}} \pm \sigma$	$T_{\text{opt}} \pm \sigma$	σ_n	σ_T
0.1	0.80 ± 0.007	0.741 ± 0.006	0.84 ± 0.015	0.702 ± 0.012	0.044	0.051
0.3	0.90 ± 0.009	0.652 ± 0.007	0.91 ± 0.017	0.624 ± 0.010	0.015	0.042
0.5	1.02 ± 0.012	0.576 ± 0.008	1.09 ± 0.025	0.550 ± 0.012	0.066	0.044
0.7	1.19 ± 0.015	0.494 ± 0.005	1.21 ± 0.029	0.476 ± 0.010	0.018	0.035
0.9	1.45 ± 0.020	0.404 ± 0.005	1.43 ± 0.036	0.402 ± 0.009	0.011	0.006
0.95	1.54 ± 0.021	0.370 ± 0.004	1.57 ± 0.041	0.371 ± 0.009	0.017	0.001

Table 6.2. Estimates of the density and derivatives with respect to the Knudsen number

x	n	σ_n	T	σ_T	n^1	σ_{n^1}	T^1	σ_{T^1}	n^2	σ_{n^2}
0.1	0.83	0.001	0.67	0.001	0.05	0.004	−0.073	0.003	−0.06	0.01
0.3	0.91	0.002	0.61	0.001	0.02	0.005	−0.045	0.004	−0.01	0.02
0.5	1.01	0.002	0.55	0.001	−0.01	0.006	−0.020	0.005	0.01	0.03
0.7	1.12	0.002	0.49	0.001	−0.05	0.008	−0.002	0.005	0.06	0.04
0.9	1.29	0.003	0.42	0.001	−0.14	0.009	0.022	0.005	0.11	0.04
0.95	1.35	0.003	0.40	0.001	−0.17	0.010	0.031	0.005	0.20	0.05

In order to check the results we perform calculations using the collision estimate. To control the derivatives obtained we calculate finite difference derivatives of the form

$$\frac{g(Kn_0 + \varepsilon) - g(Kn_0)}{\varepsilon}$$

for various ε. The results coincide to a sufficiently good precision. Let us denote by the symbol F^k the k-th order approximation to the solution to the equation

$$F^k(x, Kn) = n(x, K_0) + \sum_{j=1}^{k-1} n^i(x, Kn_0) \frac{(Kn - Kn_0)^i}{i!}.$$

Table 6.3 presents the relative errors for F^k, $k = 0, 1, 2$. The analysis of the results

Table 6.3. Relative errors of the parameter extension method

Kn	F^0	F^1	F^2
0.5	0.14	0.0699	0.0478
0.2	0.265	0.168	0.119

obtained shows that one can use the derivatives for the approximation by the parameter under small changes of the Knudsen number. It should be noted that the variance for calculating the estimates of the derivatives grows in comparison with the variance of the density and temperature estimates, which in turn naturally increases the computational cost.

The studies made show the possibility of using the given methods for solving the nonlinear integral equations. On this basis one can make a conclusion about the possibility of constructing a more efficient algorithm than a simple iteration.

Appendix A
Some Simulation algorithms

A.1. Numerical simulation of random variables

As a rule, simulation of a random variable with a given distribution is carried out by transformations of one or a number of independent values of a random number α uniformly distributed in the interval (0,1), i.e. by the formula: $\xi = \varphi(\alpha_1, ..., \alpha_n)$

1. The sequence of 'sample' values of α is usually obtained on a computer by number-theoretic algorithms, of which the most widely used is the so-called 'method of residues', in the form

$$u_0 = 1, \quad u_n \equiv u_{n-1} M (\mathrm{mod}\, 2^r), \quad \alpha_n = u_n \cdot 2^{-r}.$$

Here r is the order of the mantissa of the computer. Often $M = 5^{2p+1}$ is used [17], where

$$p = \max\{q : 5^{2q+1} < 2^r\}.$$

Numbers of this type are called 'pseudo-random numbers'; they are verified by statistical testing and by solving typical problems (see [17,24]). The length of the period of the above version of the method of residues is 2^{r-2}. Physical generators, tables of random numbers and quasi-random numbers are also used in the Monte Carlo method.

The following special order of using pseudo-random numbers is expedient to correlate different computations. It is related to conventional methods of verifying the multidimensional uniformness [17]. The sequence $\{u_n\}$ is supposed to be divided to the subsequences of the length m, beginning with the numbers u_{km}, $k = 0, 1, 2, ...$, and each subsequence is used to construct the corresponding random trajectory. Clearly,

$$u_{(k+1)m} \equiv u_{km} M^m (\mathrm{mod}\, 2^r).$$

So, to simulate the k-th trajectory we use the multiplicative pseudo-random sequence beginning with

$$\alpha_{km} = u_{km} \cdot 2^{-r}.$$

Here, it is sensible to use 'real' random numbers instead of α_{km}. This combined method has the theoretical basis provided $M \to \infty$ [17,52].

Real random numbers can be produced by physical generators. It is possible to improve their distribution by summarization modulo one, i.e., using the expression

$$\alpha = \text{fract}\left(\sum_{i=1}^{n} \alpha_i\right),$$

where $\{\alpha_i\}$ are numbers from a physical generator. It is known that the distribution of α very quickly converges to uniform distribution, if n increases. When simulating trajectories, these α are used as initial numbers for the method of residues with M as large as possible. It seems that this combined method is the most promising for solving problems considered in this monography.

During approximately 30 years the version of the method of residues with $M = 5^{17}$ and $r = 40$ was successfully used for solving different mathematical physics problems. The numerical results of the statistical testing of this version are presented in [17]. Similar positive results were obtained for the above-mentioned special order of using these numbers with $m = 1024$.

2. The standard method of simulating a discrete random variable with distribution

$$P(\xi = x_k) = p_k, \quad k = 0, 1, \ldots,$$

is as follows: put $\xi = x_m$ if the chosen value of α satisfies

$$\sum_{k=0}^{m-1} p_k \le \alpha < \sum_{k=0}^{m} p_k,$$

while m is determined by successive subtraction of the probabilities p_k from α. Clearly, the computational cost of this algorithm is proportional to the value $E\nu + 1$, where ν is the random number of the chosen value of ξ.

There is a well known special simulation formula for the uniform discrete distribution on the set $\overline{1, n}$:

$$\nu = \text{entier}(\alpha n) + 1.$$

It is possible to use this formula to improve the simulation of arbitrary discrete distributions by uniform dividing the probabilities with repeating the corresponding values.

3. The standard method of simulating a continuous random variable (sometimes called the method of inverse functions) is to use the easily verified representation

$$\xi = F^{-1}(\alpha),$$

where f is the distribution function with a given density f.

4. Sometimes, randomization of the simulation is useful (in other words, the method of superposition), based on the expression:

$$f(x) = \sum_k p_k f_k(x);$$

here one first chooses a number m with distribution $P\{m = k\} = p_k$, and then obtains a sample value ξ from the distribution with the density f_m. Accordingly to the standard simulation formula here we have the representation:

$$\xi = F_i^{-1}(\alpha), \quad \text{if} \quad S_{i-1} \le \alpha_1 \le S_i,$$

where $S_i = \sum_{k=1}^{i} p_k$. Clearly, the random variable

$$\alpha' = (\alpha_1 - S_{i-1})/p_i, \quad \text{if} \quad S_{i-1} \le \alpha_1 < S_i,$$

is independent of α_1. Hence, we can use the formula:

$$\xi = F_i^{-1}((\alpha_1 - S_{i-1})/p_i), \quad \text{if} \quad S_{i-1} \le \alpha_1 < S_i.$$

It was shown in [17], that the formula with an arbitrary set $\{p_i, F_i\}$, provided $\sum p_i F_i \equiv F$, represents the class of simulating formulas $\xi = \varphi(\alpha_1)$ with piecewise monotonous and smooth functions φ.

Now consider the special superposition simulation algorithm for the density function, determined by the linear interpolation with respect to the parameter:

$$f(x, \lambda) = \frac{\lambda - \lambda_{i-1}}{\lambda_i - \lambda_{i-1}} f_i(x) + \frac{\lambda_i - \lambda}{\lambda_i - \lambda_{i-1}} f_{i-1}(x)$$

provided $\lambda_{i-1} \le \lambda < \lambda_i$. Clearly, here it is possible to simulate ξ in accordance with $f_i(x)$, if $\alpha_1 \le (\lambda - \lambda_{i-1})/(\lambda_i - \lambda_{i-1})$, and in accordance with $f_{i-1}(x)$ - otherwise.

5. Another general method of simulating a continuous random variable is the 'rejection method' [21,24,39].

Let functions $g(x)$ and $g_1(x)$ be such, that $g_1(x) \ge g(x) \ge 0$,

$$G_1 = \int_{-\infty}^{+\infty} g_1(x)dx < +\infty,$$

and there exists a sufficiently effective simulation method for the density function $f_1(x) = g_1(x)/G_1$. The corresponding rejection algorithm for the density function $f(x) = g(x)/G$ is as follows:

1) ξ_0 is sampled from the distribution with the density $f_1(x)$ and $\eta = \alpha g_1(\xi_0)$;

2) if $\eta \le g(\xi_0)$, then the point is rejected and the simulation process continues beginning with item (1), otherwise $\xi = \xi_0$.

Clearly, the average number of operations in the rejection method is proportional to

$$S = G_1/G.$$

For example, if

$$g(x) = cf(x) \le R, \quad a \le x \le b,$$

one can take

$$\xi_0 = a + (b - a)\alpha_1, \quad \eta = R\alpha_2.$$

Here

$$S = R(b - a)/G.$$

In particular, this version of the rejection method is convenient to sample the new energy e of compton-scattered gamma-quantum with respect to the probability density function

$$f(e,E') = c\Big[\frac{e}{E'} + \frac{E'}{e} + \Big(\frac{1}{E'} - \frac{1}{e}\Big)\Big(2 + \frac{1}{E'} - \frac{1}{e}\Big)\Big],$$
$$E'(1+2E')^{-1} < e < E'.$$

Here
$$f(e,E')/c \le 1 + 2E' + (1+2E')^{-1}.$$

For solving elliptic equations by the Monte Carlo method (see Section 3.2) sometimes it is necessary to simulate the random variable with the density function, which is proportional to
$$g(r) = \begin{cases} r\frac{sh(a[R-r])}{sh(a,R)}, & r \le R \\ 0, & r > R. \end{cases}$$

Here a, R are the positive parameters. It is easy to show, that
$$g(r) \le r\exp(-ar) = g_1(r), \quad r \le R.$$

Hence, it is possible to use here the following rejection algorithm:
1) $\xi_0 = -\ln(\alpha,\alpha_2)/a$; if $\xi_0 > R$ then go to beginning, otherwise $\eta = \alpha_0 g_1(\xi_0)$;
2) if $\eta \ge g$, then item 1 is repeated and so on, otherwise $\xi = \xi_0$.
For this algorithm
$$q = 1/S = 1 - \frac{aR}{sh(aR)} \to_{aR\to\infty} 1.$$

If aR is small, then it is expedient to simulate the auxiliary point in Section 3.2 in the same way as for the Poisson equation.

6. Besides the standard simulation algorithms there exist some special simulating formulas for different distributions.

For instance, it is well known [21,24] that the expressions
$$\eta_1 = \sqrt{-2\ln\alpha_1}\cos(2\pi\alpha_2), \quad \eta_2 = \sqrt{-2\ln\alpha_1}\sin(2\pi\alpha_2)$$
represent two independent standard Gaussian random variables, so that $\mathrm{E}\eta_i = 0$ and $\mathrm{D}\eta_i = 1$, $i = 1,2$.

Note, that sometimes it is possible to construct special simulating expressions on the basis of Theorems 1,2 from [39, Appendix 2]. Examples of those expressions are represented in Table A.1.

Table A.1. Special simulating expressions

Density function	simulating expressions
$-\ln x,\ 0 < x \le 1$	$\alpha\alpha_1$
$\arccos x,\ 0 \le x \le 1$	$\alpha\sqrt{1-\alpha_1^2}$
$(n+1)(1-x^{1/n}),\ 0 \le x \le 1$	$\alpha\alpha^{n/(n+1)}$
$ce^{-\frac{x+\omega}{T}}\mathrm{sh}\frac{2\sqrt{\omega x}}{T},\ x \ge 0$	$T[(\eta+\sqrt{2\omega/T})^2/2 - \ln\alpha]$

Now consider the simulation of 'beta'-distribution with the density function

$$f_{p,m}(x) = c_{p,m}x^{p-1}(1-x)^{m-1}, \quad 0 < x < 1,$$

where $p > 0$, $m > 0$ are parameters.

In [17], there are represented the following results. If m is integer, then there exists the simulating formula

$$\xi_{p,m} = \prod_{k=1}^{m} \alpha_k^{1/(p+k-1)} = \exp\left(\sum_{k=1}^{m} \frac{\ln \alpha_k}{p+k-1}\right).$$

If p is integer, then the transformation $\xi = 1 - \zeta$ may be useful. In the case when both p and m are not integers, we can use the formula

$$\xi_{p,m} = \exp\left(\sum_{k=1}^{[m]+1} \frac{\ln \alpha_k}{\nu + p + k - 1}\right),$$

where $[m] = \text{entier}(m)$, while ν is integer random variable with the probabilities

$$p_k = \frac{[m]!}{B(p,m)} \frac{a(a+1)...(a+k-1)}{k!(k+p)(k+p+1)...(k+p+[m])},$$

where $a = [m+1] - m$. Note, that $M\nu = +\infty$ for $[m] = 0$, and the computational cost of the corresponding standard simulation algorithm is infinite. However, in the case $p < 1$ and $m < 1$ it is possible to additionally use the superposition method on the basis of the relation

$$f_{p,m}(x) = \frac{p}{p+m} f_{p+1,m}(x) + \frac{m}{p+m} f_{p,m+1}(x).$$

Now let ξ_ν be the 'gamma'-distributed random variable with the density function

$$f_\nu(x) = \frac{x^{\nu-1}e^{-x}}{\Gamma(\nu)}, \quad x > 0.$$

The following representation is well-known (see, for example, [3]):

$$\xi_\nu = -\ln\left(\prod_{k=1}^{[\nu]} \alpha_k\right) + \xi_{\nu-[\nu]}$$

Therefore, it is necessary to construct a simulating algorithm for $\nu < 1$. It this case, it is possible to use the rejection method based on the inequality

$$x^{\nu-1}e^{-x} \le g_1(x) = \begin{cases} x^{\nu-1}, & 0 < x < 1 \\ e^{-x}, & x \ge 0. \end{cases}$$

The random variable with the density function $cg_1(x)$ is easily simulated by the standard method. Here

$$S = \frac{\nu^{-1} + e^{-1}}{\Gamma(\nu)} \xrightarrow[\nu \to 0]{} 1.$$

If $\nu = 1$, then $S = 1 + e^{-1}$.

7. In the clear way the superposition and rejection methods are generalized to the simulation of random vectors. There exists a special method of simulating the Gaussian random vector ξ with the average $m = (m_1, ..., m_n)$ and the covariance matrix $K = \{K_{ij}\}$ $i,j = 1, ..., n$. This method is based on the representation

$$\xi = A\eta + m,$$

where A is the lower-triangular matrix, while η is the vector composed of the independent standard normal variables. It is possible to compute the coefficients of A by the recurrent formula

$$a_{ij} = \frac{K_{ij} - \sum\limits_{k=1}^{j-1} a_{ik}a_{jk}}{\sqrt{K_{ij} - \sum\limits_{k=1}^{j-1} a_{jk}^2}}, \quad 1 \leq j \leq i \leq n.$$

Inside the square root there is the principal minor of the matrix K. Hence, this matrix must be positive, i.e. all its eigenvalues must be strictly positive.

Note that if the entries a_{ij} are calculated approximately (for example, by the Monte Carlo method), it may appear that the calculated matrix K is not positive definite. In this case the matrix K can be improved as follows. First, a transfromation matrix T is constructed such that

$$TKT^{-1} = \Lambda = \{\lambda_{ii}\}, \quad i = 1, ..., n,$$

where Λ is a diagonal matrix. Then a matrix $\Lambda^{(0)}$ is constructed by replacing in Λ the negative values λ_{ij} by $|\lambda_{ij}|$, and zero elements – by small positive values ε. The transformation gives:

$$K^{(0)} = T^{-1}\Lambda^{(0)}T.$$

In what follows, we use or a_{ij} the entries of the matrix $K^{(0)}$.

In [40], the 'repetition method' is represented for simulation of the random vector with positive covariances and a given one-dimensional distribution, the same for all components.

In this method, a set of components is randomly divided into the groups with coinciding values, which are then independently sampled from the given distribution. Evidently, for the fixed division Δ_m there is a random vector with the correlation matrix

$$R(\Delta_m) = \{r_{ij}^{(m)}\} \quad i, j = 1, ..., n,$$

such that $r_{ij} = 1$ if ξ_i, ξ_j belong to the same group and $r_{ij} = 0$ otherwise.

Therefore, for the probabilities $p_m = P(\Delta_m)$ the following equation takes place:

$$R = \sum_{m=1}^{N} p_m R(\Delta_m).$$

If the appropriate solution of this equation exists, then it is possible to use the repetition method and it is not possible otherwise (see [40]).

Note that here it is possible to improve realization, if a given one-dimensional distribution is infinitely divisible (see Section A.2), for instance, the Gaussian distribution.

A random vector with given one-dimensional distribution functions $\{F_k\}$ and with a given correlation matrix can be sometimes constructed by the formulae

$$\xi_k = F_k^{-1}[\Phi(\eta_k)], \quad k = 1, ..., n,$$

where Φ is the distribution function of the standard Gaussian random value, and $(\eta_1, ..., \eta_n)$ is a vector of standard Gaussian random values with covariances satisfying the known system if equations, It may appear that this system is not solvable and this method is not applicable.

A.2. Numerical simulation of random fields

1. Simple facts from convergence theory for random functions are firstly presented. Consider a random function (r.f.)

$$\xi(t) = \xi(t, \omega),$$

which is a random variable for any fixed value of t. The distribution function of this random variable is defined as

$$F_1(t, x) = \mathbf{P}(\xi(t) < x).$$

Information about the random functions used in Monte Carlo methods is given in [17]. Note that the definition of r.f. can be given through the set of finite dimensional distribution functions

$$F_n(t_1, x_1; ...; t_n, x_n) = \mathbf{P}\big(\xi(t_1) < x_1, ..., \xi(t_n) < x_n\big).$$

The set must satisfy so-called self-consistent conditions; for example, the marginal distribution functions corresponding to a given function F_n must belong to this set.

If t is a one-dimensional variable, then $\xi(t)$ is called a random function; if t is multidimensional variable, then $\xi(t)$ is called a random field. Next we consider some questions of the weak convergence theory for sequences of r.f. Notice that the weak convergence of finite dimensional distributions is defined here as the convergence of $F_n^{(m)}$ to F_n as $m \to \infty$, $n = 1, 2, ...$ at all points of continuity of F_n [3].

Assume that in the space of samples of r.f. $\xi = \xi(t)$ a metric L is introduced. Consider a set Φ of functionals $\varphi(\xi)$ which are continuous in this metric: $\varphi \in \Phi$. We shall say that the sequence $\xi_n(t)$ is weakly convergent to $\xi(t)$ in the metric L (this will be denoted by $\xi_n \Rightarrow \xi$) if the distribution function of the random variable $\varphi(\xi_n)$ weakly converges to the distribution function of $\varphi(\xi)$ for all $\varphi \in \phi$.

Suppose now that $t \in T$ is a bounded closed set, and the samples of $\xi(t, \omega)$ are continuous (for almost all ω). Then in the sample space of ξ it is natural to introduce a metric as follows:

$$\rho(\xi_1, \xi_2) = \max_t |\xi_1(t) - \xi_2(t)|. \tag{A.1}$$

The following statement is in fact the known [3] Kolmogorov criterion for slow convergence of continuous r.f.

Lemma A.1. *If the finite dimensional distributions of the processes $\xi_n(t)$ weakly converge to finite dimensional distributions of $\xi(t)$, and for arbitrary t_1, t_2 and some constants $c, \gamma, \alpha > 0$ the following inequality holds*

$$\mathrm{E}|\xi_n(t_2) - \xi_n(t_1)|^\gamma \le c|t_2 - t_1|^{1+\alpha}$$

then $\xi_n \Rightarrow \xi$ in the metric (A.1). □

When random functions with discontinuous realizations are considered (i.e. the samples of these r.f. can be discontinuous with non-zero probability), special functional space D are introduced [3]. Let $t \in [0, 1]$; then D is defined as a set of real valued functions $x(t)$, such that at an arbitrary point the left and the right limits exist. It is also assumed that

$$x(t) = x(t+0), \quad x(0) = x(+0), \quad x(1) = x(1-0).$$

It is known that in the uniform metric (A.1) the space D is not separable. Therefore in D a special metric which is slightly weakened compared to the uniform metric is introduced.

Denote by Λ a set of continuous, monotonically increasing on [0,1], real valued functions λ such that $\lambda(0) = 0$, $\lambda(1) = 1$. The metric in D is defined by

$$\rho_D(x,y) = \inf_{\lambda\in\Lambda} \left[\sup_t |x(t) - y(\lambda(t))| + \sup_t |t - \lambda(t)|\right]. \tag{A.2}$$

If $x(t)$ and $y(t)$ are continuous then

$$\rho_D(x,y) = \rho(x,y).$$

The space D in the metric ρ_D is separable [3].

The next proposition is the *Kolmogorov-Chentsov* slow convergence criterion on the metric (A.2).

Lemma A.2. *If the finite dimensional distributions of the processes $\xi_n(t)$ without discontinuities of second order weakly converge to the finite dimensional distributions of the process $\xi(t)$ and for arbitrary $0 \le t_1 < t_2 < t_3 \le 1$, the following inequality holds*

$$\mathrm{E}\left[|\xi(t_3) - \xi(t_2)|\,|\xi(t_2) - \xi(t_1)|\right]^{\gamma} \le (t_3 - t_1)^{1+\alpha}$$

for some $c, \gamma, \alpha > 0$, then $\xi_n \Rightarrow \xi$ with respect to the metric (A.2). □

2. A simple 'randomized' method of numerical construction of a homogeneous Gaussian random field with a given spectral density is now presented. This method is determined by the corresponding randomized representation of the spectral decomposition of the correlation function.

Let $\xi(x)$ be a random real field with a correlation function $B(x)$ which possesses the spectral decomposition

$$B(x) = \int \cos(\lambda x) p(\lambda) d\lambda, \tag{A.3}$$

where $p(\lambda)$ is the spectral density [19] (here $x, \lambda \in R_n$). Note that representation of Eq. (A.3) exists in particular if

$$\int |B(x)| dx < +\infty.$$

To simplify the exposition we shall assume that

$$\mathrm{E}\xi(x) = 0, \quad \mathrm{D}\xi(x) = 1.$$

There are available methods of numerical simulation (see, e.g., [17]) of Gaussian random processes and fields with rational spectral densities; these methods are based on linear transformations of 'white noise' with suitable spectral characteristics. These methods may be too labor-consuming, since they require solutions of differential equations (in the case of a field-partial differential equations). In this section we present not a completely adequate (in the sense of finite-dimensional distributions), but a very simple method of numerical construction of a random field with an arbitrary spectral density. For this purpose we use in particular the well-known fact that in the case of a finite spectrum, i.e. when

$$B(x) = \sum_{k=1}^{m} p_k \cos(\lambda_k x),$$

the required field can be constructed using the formula

$$\xi(x) = \sum_{k=1}^{m} \sqrt{p_k}[\xi_k \sin(\lambda_k x) + \eta_k \cos(\lambda_k x)], \tag{A.4}$$

where (ξ_k, η_k), $k = 1, 2, ..., m$, are jointly independent standard Gaussian random variables.

Representation of Eq. (A.4) can be utilized in an obvious manner for the approximate construction of a field with a continuous spectrum based on a quadrature formula of the type

$$B(x) = \int \cos(\lambda x) p(\lambda) d\lambda \approx \sum_{k=1}^{m} p_k \cos(\lambda_k x) \tag{A.5}$$

with nonnegative coefficients which add up to one.

A formula of the form Eq. (A.5) can sometimes be 'randomized' in such a manner that it yields an unbiased estimator of the integral. For our purpose, this fact may be utilized, based on following assertion which shows the connection between the 'randomization' of a correlation function and the 'randomization' of a field.

Lemma A.3. *Let a homogeneous field $\Psi(x)$ be stochastically associated with a random vector χ having the distribution $P_\chi(dy)$.*

Then the correlation function $B(x)$ of the field Ψ can be written as

$$B(x) = \int B(x, y) P_\chi(dy) = \mathrm{E} B(x, \chi), \tag{A.6}$$

where $B(x, y)$ is the conditional correlation function of the field $\Psi(x)$, given $\chi = y$. □

This lemma is a direct corollary of the elementary 'formula of the total mathematical expectation'. It shows that one can construct a field with a correlation function of Eq. (A.6) in two stages: 1) choose a value of the vector χ; 2) construct a realization of the 'randomized' field with the correlation function $B(x, \chi)$.

Subdivide the space $\Lambda = R_n$ into parts $\Lambda_1, ..., \Lambda_m$ and let the random points $\lambda_1, ..., \lambda_m$ be distributed in these parts according to the densities

$$p_k(\lambda) = p(\lambda) \Big/ \int_{\Lambda_k} p(\lambda) d\lambda, \quad \lambda \in \Lambda_k.$$

Then in view of Eq. (A.3) we have

$$B(x) = \mathrm{E} \sum_{k=1}^{m} p_k \cos(\lambda_k x), \quad p_k = \int_{\Lambda_k} p(x) dx.$$

Hence on the basis of the previous discussion we obtain the following algorithm: 1) choose (dependently or independently) values for $\lambda_1, ..., \lambda_m$; 2) construct a realization of the field using formula Eq. (A.4) with the determined values of p_k, λ_k, $k = 1, ..., m$.

In this manner a 'randomized' random field of the form of Eq. (A.4) with a given spectral density is constructed. It is well known (see, e.g., [17], p.42) that the values for ξ_k and η_k can be computed from the formulas

$$\xi = \sqrt{-2\ln\alpha'}\cdot\cos 2\pi\alpha, \quad \eta = \sqrt{-2\ln\alpha'}\cdot\sin 2\pi\alpha,$$

where α' and α are independent random variables, uniformly distributed on $(0,1)$. The corresponding expression Eq. (A.4) can be written as

$$\xi_m(x) = \sum_{k=1}^{m}\sqrt{-p_k\ln\alpha'_k}\cdot\cos(\lambda_k x - 2\pi\alpha_k), \tag{A.7}$$

where the λ_k are random and distributed as indicated above.

The conditional one-dimensional distribution of the field Eq. (A.7) for fixed values of $\lambda_1, ..., \lambda_m$ is the standard Gaussian distribution; consequently the marginal one-dimensional distribution of Eq. (A.7) is also standard Gaussian. Analogous considerations show that the multidimensional distributions of the field Eq. (A.7) are not normal and that the field is not ergodic, but these defects diminish as $m \to \infty$ and as the subdivision of space R_n becomes uniformly more and more 'refined'. It is not hard to see that the field under consideration is homogeneous.

In applications we often encounter problems in which the properties of a solution are completely determined, with required precision, by the correlation function of the initial field (such is the case, for example, with the problem of estimating the mathematical expectation of a solution whose dependence on the field is approximately quadratic, within the limits of statistical fluctuations). In this case one may assume that $m = 1$, $\Lambda_1 = R_n$, and representation Eq. (A.7) becomes especially simple. Below we present an example of such a representation.

Example. Consider a two-dimensional field $\Psi(x, y)$ with a correlation function

$$B(r) = k_0 r K_1(k_0 r),$$

where K_1 is the Macdonald function [51], k_0 is a constant and $r^2 = x^2 + y^2$. The field of the stream function has a correlation function of this type under the hypothesis that the distribution of the potential vortex is haotic. The corresponding spectral density is equal to

$$p(\mu, \nu) = (\lambda^2 + k^2)^{-2}$$

where $\lambda^2 = \mu^2 + \nu^2$. Passing to polar coordinates, one can easily obtain the simulation formulas for the distribution with density $p(\mu, \nu)$:

$$\mu = k_0\sqrt{\frac{\alpha_1}{1-\alpha_1}}\cos 2\pi\alpha_2, \quad \nu = k_0\sqrt{\frac{\alpha_1}{1-\alpha_1}}\sin 2\pi\alpha_2.$$

Hence we obtain the following representation for Ψ of the form Eq. (A.7):

$$\Psi(x, y) = \sqrt{-2\ln\alpha'}\cos\left[k_0\sqrt{\frac{\alpha_1}{1-\alpha_1}}(x\cos 2\pi\alpha_2 + y\sin 2\pi\alpha_2) + 2\pi\alpha'\right],$$

where $\alpha_1, \alpha_2, \alpha', \alpha'$ are independent and uniformly distributed on $(0, 1)$.

We also note that for simulation of correlation properties of a field one can utilize simpler variables ξ and η, for example those which take on values ± 1 with equal probabilities; in this case one can more easily ensure the nonnegativity of realizations which may be required. All of what has been said above can be extended in an obvious manner to the case of a discrete spectrum.

Now consider the convergence of $\xi_m(x)$ to the normal field with a given spectral density $p(\lambda)$. As mentioned above, this convergence is connected with the partitioning of Λ into subspaces Λ_k. Let V_k be the value of Λ_k, and let d_k be the diameter of Λ_k. We shall use uniform (with respect to the volume) partitions of the domain $\Lambda \setminus \Lambda_m$ such that

$$d_k \leq c V_k^{1/n}, \quad k = 1, 2, ..., m - 1,$$

where $V_k \to 0$ as $m \to \infty$. Such partition can be obtained, for example, by using a uniform rectangular mesh with a step size $m^{-1/n}$.

Theorem A.1. *Let $\xi_m(x)$ be determined by formula Eq. (A.7) for $|x| < R, \Lambda_m = \{\lambda : |\lambda| > l_m\}$ and $d_k < c l_m m^{-1/n}$, $k = 1, 2, ..., m - 1$. Then there is a Gaussian field $\xi(x)$ with the spectral density $p(\lambda)$ so, that*

(1)

$$\mathrm{E} \int_{|x|<R} [\xi_m(x) - \xi(x)]^2 dx \to 0$$

if $l_m m^{-1/n} \to 0$ and $l_m \to \infty$ as $m \to \infty$; and

(2)

$$\mathrm{E} \int_{|x|<R} [\xi_m(x) - \xi(x)]^2 dx \leq c m^{-2\varepsilon[n(2+\varepsilon)]}$$

for $l_m = c m^{2/[n(2+\varepsilon)]}$ if, for some positive ε, $\int |\lambda|^{\varepsilon} p(\lambda) d\lambda < +\infty$.

Theorem was proved in [40]. □

Note that the estimate is optimal (by order of magnitude) if l_m is determined from

$$P_m = \int_{|\lambda|>l_m} p(\lambda) d\lambda = l_m^2 m^{-2/n}.$$

In addition, if $p(\lambda)$ decreases rapidly enough (for example, exponentially) as $|\lambda| \to \infty$, then P_m is bounded by the quantity $c^* m^{-2/n} L(m)$, where $L(m)$ is a slowly increasing function. The convergence (in probability) of the L_2-error $\varepsilon = \|\xi_n(x) - \xi(x)\|_{L_2}$ to zero, where $\xi(x)$ is a normal random field with a given density, thus follows by the Chebyshev inequality. Similar statements could be formulated and proved for the convergence of the derivatives of $\xi_n(x)$ under the condition that there exist appropriate moments of the spectral density. For example,

$$\mathrm{E} \int [\xi_m^{(k)}(x) - \xi^{(k)}(x)]^2 dx \leq c m^{-2\varepsilon/[n(2+\varepsilon)]} \tag{A.8}$$

if

$$\int |\lambda|^{2+\epsilon} p(\lambda) d\lambda < +\infty \tag{A.9}$$

where $\xi^{(k)}$ is the partial derivative of ξ with respect to $\lambda^{(k)}$, $k = 1, 2, ..., n$.

Note that sometimes it is possible to prove the convergence (in probability) to zero of uniform deviation of ξ_m from ξ using imbedding theorems. In particular, for random processes $\xi = \xi(t)$ (i.e. for $n = 1$) it follows from Eq. (A.8), (A.9) and by imbedding theorem ($W_2^{(1)}$ in C) that

$$\mathrm{E} \sup_t |\xi_m(t) - \xi(t)| \leq c_0 m^{-\epsilon[n(2+\epsilon)]},$$

Note that the estimates obtained can be used to investigate the convergence of the functionals studied, since convergence in probability implies here weak convergence with respect to the appropriate metric (see, for example, [3]).

Now let us consider similar approximate modes of the vector homogeneous Gaussian random fields, developed in detail in [47].

Let $V(x) = (v_1(x), ..., v_l(x))^T$ be a vector, generally speaking, complex-valued homogeneous field with a given spectral tensor

$$F(\lambda) = \frac{1}{(2\pi)^n} \int_{R_n} B(\rho) e^{-i(\lambda,\rho)} d\rho$$

where $B(\rho) =< V(x+\rho) V^*(x) >$ is a covariance tensor of the field $V(x)$. Let $p(\lambda)$ be a probability density function in R_n. The simplest simulation formula (for $m = 1$) here is:

$$V(x) = \frac{1}{\sqrt{p(\lambda)}} e^{i(\lambda,x)} \zeta_\lambda,$$

where the vector λ is distributed with the density function $p(\lambda)$, and for a given λ, ζ_λ is the complex Gaussian random l-dimensional vector-column with zero average and the following covariance matrix

$$< \zeta_\lambda \zeta_\lambda^* >= F(\lambda).$$

It is possible to construct the vector ζ_λ in the following conventional way. Let $F(\lambda) = G(\lambda) G^*(\lambda)$, where $G(\lambda)$ is a lower-triangular matrix. Then

$$\zeta_\lambda = G(\lambda) \xi_\lambda,$$

where ξ_λ is a l-dimensional standard Gaussian random vector with zero average and unity covariance matrix. It is naturally to choose $p(\lambda)$ so that:

$$p(\lambda) = \frac{1}{u_*^2} S_p F(\lambda) = \frac{1}{u_*^2} \sum_{i=1}^{l} F_i(\lambda),$$

where $u_*^2 = \int_{R^n} S_p F(\lambda) d\lambda$ is the total field energy.

3. Now let us consider the case of a real vector random field $V(x)$. Generalization of the simplest (for $m = 1$) formula of Eq. (A.7) in this case is determined by the expression:

$$V(x) = \frac{1}{\sqrt{p(\lambda)}}(\zeta_\lambda \cos(\lambda, x) + \eta_\lambda \sin(\lambda, x)).$$

Here a random vector λ is distributed with a density function, and for a given λ, the $2l$-dimensional real Gaussian random vector-column $(\zeta_\lambda, \eta_\lambda)^T$ has zero average and the covariance matrix

$$H = \begin{pmatrix} \mathrm{Re}F(\lambda) & \mathrm{Im}F(\lambda) \\ -\mathrm{Im}F(\lambda) & \mathrm{Re}F(\lambda) \end{pmatrix}$$

It is also possible to simulate this Gaussian vector in the above-mentioned way, related to the triangular factorization of the matrix H.

However it is possible to use in this case a simpler triangular factorization $F(\lambda) = GG^*$, because the following representations hold:

$$H = \begin{pmatrix} \mathrm{Re}G & \mathrm{Im}G \\ -\mathrm{Im}G & \mathrm{Re}G \end{pmatrix} \begin{pmatrix} \mathrm{Re}G & \mathrm{Im}G \\ -\mathrm{Im}G & \mathrm{Re}G \end{pmatrix}^T$$

and

$$\begin{pmatrix} \xi_\lambda \\ \eta_\lambda \end{pmatrix} = \begin{pmatrix} \mathrm{Re}G & \mathrm{Im}G \\ -\mathrm{Im}G & \mathrm{Re}G \end{pmatrix} \begin{pmatrix} \xi'_\lambda \\ \eta'_\lambda \end{pmatrix},$$

where $\xi'_\lambda, \eta'_\lambda$ are independent l-dimensional standard Gaussian vectors.

Consider now spectral models of non-Gaussian random fields related to stationary point fluxes.

To solve stochastic problems by the method of statistical modelling it is necessary to construct numerically the realizations of the random processes and fields. It is then often important to preserve the correlation function and the one-dimensional distributions, all the more so because in the case of non-Gaussian fields there is no satisfactory information about multi-dimensional distributions. However, it is often desirable to have a set of models of random fields; on the basis of this set of models, it is possible to study the sensitivity of functionals to the variations of multi-dimensional distributions when they converge to distributions of a random field with appropriate (for example, continuous) realizations. Models of this kind which are sufficiently close (in the sense of weak convergence) to the limit fields and realizable on a computer are naturally used as approximate models. The more accurate are these approximations, the larger is the computational cost of realization.

4. Here we consider special random processes which are connected with stationary fluxes of points:

$$\tau_k = \sum_{i=1}^{k} \eta_i, \quad k = 1, 2, ...; \quad \tau_0 = 0, \tag{A.10}$$

where η_i are independent non-negative random variables with distribution densities $f_i(x)$, $i = 1, 2, \ldots$.

These processes are called Palm fluxes [19]. The following relations hold for them:

$$f_k(x) = f(x) = F'(x), \quad k = 2, 3, ...; \quad f_1(x) = \mu^{-1}[1 - F(x)]. \tag{A.11}$$

It is assumed here that

$$\mu = \int x f(x) dx < +\infty.$$

On the other hand, it follows from Eq. (A.11) that the flux τ_k is stationary (see, for example, Chapter XI in [19]).

We use fluxes of form Eq. (A.10) with a given probability $p_0(t) = P(k = 0; t)$ on the interval $[0, t]$.

The Palm flux is known to be fully defined by $p_0(t)$ [19] such that

$$f(x) = -p_0''(x)/p_0'(0), \quad f_1(x) = -p_0'(x), \quad x \geq 0. \tag{A.12}$$

It directly follows from Eq. (A.11) that

$$p_0(t) = \frac{1}{\mu} \int_t^{\infty} [1 - F(t')] dt'.$$

Therefore,

$$p_0''(t) \geq 0, \quad |p_0'(0)| < +\infty, \quad p_0(+\infty) = 0. \tag{A.13}$$

It is not difficult to show that Eq. (A.13) implies inequality $p_0' \leq 0$.

Suppose that a given non-negative function $p_0(t)$, $t \geq 0$, satisfies condition of Eq. (A.13). It is then not difficult to verify that functions of Eq. (A.12) are the distribution densities for which the second condition in Eq. (A.11) holds. We thus have

Lemma A.4. *Let $p_0(t)$ be a function such that*

$$p_0(0) = 1, \quad p_0(+\infty) = 0, \quad |p_0'(0)| < \infty$$

and

$$p_0''(t) \geq 0 \quad \text{for} \quad t \geq 0.$$

If Eq. (A.11), Eq. (A.12) hold for flux Eq. (A.10), then

$$P(k = 0; t) = p_0(t).$$

□

Using the Palm flux $\{\tau_k\}$ with $P(k = 0; t) = p_0(t)$, now construct a stationary random process $\xi(t)$, $0 < t < T$ with a given one-dimensional distribution $F_\xi(x)$ as follows:

(1) A sequence $\{\tau_k\}$ is simulated until the first exit outside $(0, T)$.

(2) One takes $\xi(t) \equiv \xi_i$ in each interval (τ_{i-1}, τ_i), $i = 1, 2, \ldots$, where $\{\xi_i\}$ are independent random variables with the distribution function $F_\xi(x)$.

Denote this process by $\xi(t; p_0)$. The standardized correlation function of the process $\xi(t; p_0)$ is equal to $p_0(t)$ since

$$E[\xi(t'; p_0)\xi(t' + t; p_0)] = p_0(t) M\xi^2 + [1 - p_0(t)](M\xi)^2.$$

It is assumed here that $M\xi^2 < +\infty$. We have thus constructed a stationary process with a given one-dimensional distribution function $F_\xi(x)$ and a convex standardized correlation function $k(t) = p_0(t)$. Note that in accordance with the standard method of 'inverse distribution function' (see, for example, [24]), the realization of τ_k can be constructed by solving the equalities

$$k(\eta_1) = \alpha_1, \quad k'(\eta_i) \equiv \alpha_i k'(0), \quad i = 2, 3, \dots,$$

where $\{\alpha_i\}$ are independent random variables uniformly distributed on $(0, 1)$.

Consider the case of monotonic function $k(t)$ which can be represented by

$$k(t) = \int_0^\infty e^{-\lambda t} \mu(d\lambda),$$

where μ is a probabilistic measure on $(0, \infty)$ such that

$$P(\lambda = 0) = 0 \quad \text{and} \quad \int \lambda \mu(d\lambda) < +\infty.$$

In this case, Eq. (A.13) holds for $p_0(t) = k(t)$ and

$$f(x) = \int_0^\infty \lambda^2 e^{-\lambda x} \mu(d\lambda) \left[\int_0^\infty \lambda \mu(d\lambda) \right]^{-1}, \quad f_1(x) = \int_0^\infty \lambda e^{-\lambda x} \mu(d\lambda).$$

If there exist convenient methods of sampling the measures μ and $\lambda\mu$, then it is useful to apply the superposition method (see, for example, [24]) to sample η_i.

Note that there exists a randomized method for constructing a process with an absolutely monotone correlation function $k(t)$. Indeed, one samples λ with respect to μ and then evaluates $\xi(t; e^{-\lambda t})$.

The process constructed in this manner is obviously not ergodic while the process $\xi(t; k)$ considered above is ergodic and therefore more natural and important for applications.

Let D be a domain of n-dimensional Euclidean space lying in a cube $[0, T]^n$. Construct a random field $\xi(r)$, $r = (x_1, \dots, x_n)$ in D as follows:

(1) Sample a Palm flux $\{\tau_k^{(i)}\}$ with $p_0^{(i)}(t) = k_i(t)$ on the ith axis.

(2) Sample an independent value of $\xi(r)$ according to the distribution function $F_\xi(x)$ for each parallelepiped of the form

$$[\tau_{k_1-1}^{(1)}, \tau_{k_1}^{(1)}] \cdots [\tau_{k_n-1}^{(n)}, \tau_{k_n}^{(n)}].$$

Denote this field by $\xi(r; k)$, where $k = (k_1, \dots, k_n)$. The field $\xi(r; k)$ is homogeneous and has a standardized correlation function

$$K(r; k) = \Pi_{i=1}^n k_i(|x_i|), \quad r = (x_1, \dots, x_n). \tag{A.14}$$

Note that the set of simulated correlation functions can be extended, replacing x_i in Eq. (A.14) by $\lambda_i x_i$, where $\lambda = (\lambda_1, \dots, \lambda_n)$ is chosen at random according to the probabilistic measure $\mu(d\lambda)$ defined on $[0, +\infty]^n$ for every realization of the field. In this manner it is possible to simulate fields with the correlation function

$$K(r) = \int \exp(-\lambda_1 |x_1| - \dots - \lambda_n |x_n|) \mu(d\lambda).$$

To simulate an isotropic random field, we use the following obvious technique. First, one constructs a realization of the field $\xi(r; k)$, then rotates it isotropically and orients it according to a null vector $\omega = (\omega_1, \dots, \omega_n)$. When the correlation function of such a field is being evaluated, it is possible to consider the orientation as a fixed one, and r as an isotropic vector, i.e. $r = t\omega$, $\quad t = |r|$. Therefore,

$$K(t) = \int k_1(|\omega_1|t) \cdots k_n(|\omega_n|t) d\omega.$$

The simplest model of this kind is constructed as follows:

(1) A Palm flux $\{\tau_k\}$ with $p_0(t) = k(t)$ is sampled along an isotropic direction ω inside the domain D.

(2) An independent value of ξ is sampled according to $F_\xi(x)$ for each layer between parallel planes, perpendicular to the direction ω, constructed at points τ_k.

We call such a field model an 'isotropic-layer' model. We have for this model

$$K(t) = c_n \int_0^{\pi/2} k(t \cos \vartheta) \sin^{n-2} \vartheta d\vartheta.$$

For example,

$$K(t) = \int_0^1 \int_\Lambda e^{-\lambda t \nu} \mu(d\lambda) d\nu = \int_\Lambda \frac{1 - e^{-\lambda t}}{\lambda t} \mu(d\lambda),$$

if k is absolutely monotonic and $n = 3$. Distribution of the isotropic-layer field is not natural in some applications; it can sometimes be improved by the simple modification considered below.

Assume the distribution of ξ to be infinitely divisible, i.e. the representation

$$\xi = \xi_1^{(m)} + \cdots + \xi_m^{(m)}$$

holds for every positive integer m, where $\xi_i^{(m)}$ are independently and identically distributed according to the distribution function $F_m(x)$.

Consider the process

$$\zeta_m(t) = \sum_{i=1}^{m} \xi_i^{(m)}(t; k), \tag{A.15}$$

where $\xi_i^{(m)}(t; k)$ are independent realizations of the process $\xi^{(m)}(t; k)$ distributed according to the one-dimensional distribution $F_m(x)$. Obviously, the one-dimensional distribution function of $\zeta_m(t)$ is equal to $F_\xi(x)$ and the correlation function is equal to $k(t)$.

The realizations of the process $\zeta_m(t)$ are improved [as compared with $\xi(t; k)$] in the sense that they are closer to continuous functions, because $\zeta_m(t)$ takes constant values in smaller domains and these values are dependent. A weak convergence of the finite-dimensional distributions corresponding to $\xi_m(t)$ to distributions satisfying the consistency condition is obtained in [40]. The Chentsov-Kolmogorov criterion [6] provides the validity for the following theorem.

Theorem A.2. *Suppose that $|k''(t)| < c < \infty$ on $[0, T]$ and $\xi \geq 0$. Then the processes $\xi_m(t)$ are weakly (with respect to the metric of D) convergent to a process which is determined by the limit finite-dimensional distributions with the correlation function $k(t)$ and one-dimensional distribution function $F_\xi(x)$.*

This theorem does not include the case of Gaussian distribution. However, the convergence of $\zeta_m(t)$ is then easily obtained from the following known generalization of the mentioned criterion:

$$P\{|\zeta_m(t_3) - \zeta_m(t_2)| \geq \lambda, |\zeta_m(t_2) - \zeta_m(t_1)| \geq \lambda\} \leq \lambda^{-2\gamma}(t_3 - t_1)^{2\alpha},$$

$$\gamma = \alpha = 1.$$

The considered theorem was extended to random fields applying the appropriate generalization of the Chentsov-Kolmogorov criterion [42]. In any case, the finite-dimensional distributions of the field models considered are improved by summation of the form Eq. (A.15).

The isotropic layer model thus makes it possible to construct random fields having quite natural realizations. Note that if models of random media of form Eq. (A.15) are used in the radiation transfer it is then sufficient to take m such that the mean size of the region where $\zeta_m(r)$ is constant is less that the free path length of the radiation quantum.

We mention another useful property of the models considered. It is possible to carry out a dependent simulation on a single Palm sequence corresponding to a given function $k(t)$ of fields with different one-dimensional distributions. Numerical simulation of these models shows that non-Gaussian random fields differ essentially from Gaussian fields. Indeed, non-Gaussian fields have regions of constant values such that their diameters are essentially larger than the correlation scale of the field. This fact can be interpreted as a manifestation of the entropy maximality of the Gaussian distribution when covariances are fixed.

Theorem A.2 remains to be valid [42] if instead of the above fields the next original field is used:

$$\xi^{(m)}(r, k) = \sum_{j=1}^{n} \xi_j^{(mn)}(r_j; k_j), \tag{A.16}$$

where the processes $\xi_j^{(mn)}(x_j; k_j)$ are independent. The field (A.16) has the one-dimensional distribution $F_m(x)$ and the correlation function

$$k(r) = \frac{1}{n} \sum_{j=1}^{n} k_j(x_j).$$

It is possible here to extend the class of correlation functions using the argument transformations and the averaging considered above.

The obtained results can be extended to vector fields if ξ is regarded as a random vector with a given distribution. Here the correlation function of the simulated vector field is represented by the product of the scalar function (for instance (A.14) or (A.16)) and the correlation function of the vector ξ. The correlation function class can be also extended on the basis of the argument transformations and the averaging. The theory of the weak convergence for vector fields can be constructed similarly to the scalar case, it is only necessary to change the module for proper vector norm in the expression of the weak convergence criterion [42].

It is evident that the considered limit fields are infinitely divisible.

A.3. Remarks about simulation algorithms with the use of multiprocessor systems

1. First, it is obvious that the use of M independent computational processors by distributing random samples among them decreases the cost of the Monte Carlo method by

M times, as the final summation and averaging of the results are practically inessential. When M is large, the necessary amount of random numbers is also very large, and it is especially expedient to use the combined random-pseudorandom sequence considered in the beginning of Section A.2.

Note that for global estimating a solution in the metric C by simulation of series of trajectories from different points, it is reasonable to use the same random numbers for each point (see the end of Section 1.5). The fact decreases the necessary amount of random numbers.

2. The conclusion of Section 3.6 shows that the use of a simple parallel summator can essentially raise the efficiency of the 'global walk-on-grid' method in solving difference equations. At the same time such a device supplemented with a purely arithmetic parallel processor computing the values of the kernel also can essentially reduce the cost of the general method of local estimates in solving multi-dimensional integral equations; here the local estimate method becomes essentially more effective than the 'frequency polygon' method (see section 1.5). The computer costs can be similarly decreased when solving a number of problems by using weights (see Sections 1.1 and 1.2) and when numerically simulating the random fields by summarizing the independent realizations of the initial random functions (see Section A.2)

3. Now consider the simulation of a particle free-path in the medium with a piece-wise-constant total cross-section (see section 5.1). Let a set of surfaces is determined so that the i-th surface is the boundary only of two subdomains with numbers k_i and l_i, i.e., there is the correspondence: $i \to (k_i, l_i)$, $i = 1, ..., N$. In each given subdomain a cross-section is constant. Let a particle start from a point r in the i-th subdomain in the direction ω, i.e., along the ray

$$r(t) = r + \omega t, \quad t > 0.$$

Using standard geometric algorithms (see, for instance, [28]), it is possible to calculate the corresponding distances from the point r till all the surfaces of the system under consideration. Obviously, a simple arithmetic parallel multiprocessor system is here essentially useful. As a result we obtain the sequences:

$$t_1, t_2, ..., t_n,$$
$$(k^{(1)}, l^{(1)}), (k^{(2)}, l^{(2)}), ..., (k^{(n)}, l^{(n)}),$$

where $\{t_s\}$ are above-mentioned distances in increasing order, and $\{(k^{(s)}, l^{(s)}\}$ are the numbers of subdomains which are separated by the corresponding surfaces. Further, the sequence

$$m_1 = j, m_2, ..., m_n \tag{A.17}$$

of numbers of intersected domains can be determined by the following recursive procedure:

$$\text{if} \quad m_s = k^{(s)} \quad \text{then} \quad m_{s+1} = l^{(s)} \quad \text{else} \quad m_{s+1} = k^{(s)}, s = 1, ..., n-1.$$

Note that in the case $m_{s+1} = k^{(s)}$ the equality $l^{(s)} = m^{(s)}$ has to be valid if the initial geometrical information is true. Using the sequence (A.17) we can sample a free-path length (see section 5.1); here it is useful to previously compute in the parallel manner all the values

$$\delta\tau_s = (t_s - t_{s-1})\sigma_{m_s}, \quad t_0 = 0, \quad s = 1, ..., n.$$

If the system contains a part with a regular net of cells, then it is expedient to use for this part the maximal cross-section method (see section 5.1). Such a combined simulation of a free-path length is in detail considered for the case of a hexagonal net in [41], where the net is reduced to the parallelepipedal form, and the operation 'entier' is used for the determination of the elementary subdomain numbers.

Finally note, that if

$$\sigma(r) = \sum_{i=1}^{m} \sigma_i(r),$$

then the free-path length l can be sampled by the formula:

$$l = \min(l_1, ..., l_m),$$

where $\{l_k\}$ are independent random free-path lengths corresponding to the cross-sections $\{\sigma_k(r)\}$.

4. When simulating the 'walk on spheres' process it is necessary to determine the distance from a given point r to the boundary of the region under consideration (see section (3.1). This distance is expressed by the formula

$$d(r) = \min(d_1(r), ..., d_N(r)),$$

where $\{d_i(r)\}$ are the distances form r to the elementary surfaces, which are the parts of the boundary. It is expedient to use here a simple multiprocessor system. Usually the region is artificially divided into the parts so that if a point is included in one of them, then it is necessary to compute only some corresponding distances $\{d_i(r)\}$.

To simulate 'walk on spheres' it is necessary to sample the isotropic unit vector. In n-dimensional case it is expedient to use for this purpose the following well-known relation between the Gaussian distribution and the isotropic direction: if $\eta_1, ..., \eta_n$ are standard independent Gaussian random variables, then the vector $\bar{\eta} = (\eta_1, ..., \eta_n)$ is isotropic [17].

References

[1] H. Amann, Optimale Anfangsverteilungen bei der Monte-Carlo-Methode mit Informationsspeicherung, *ZAMM* (1967) **47**, 285–299 (in German).

[2] V.S. Antyufeev and G.A. Mikhailov, Monte Carlo restoration of the altitude curve for the scattering coefficient and calculation of the correlation characteristics of the diurnal horizon brightness. *Izv. Akad. Nauk SSSR, Ser. Fiz. Atmos. Okeana* (1976) **12**, 485–493 (in Russian).

[3] P. Billingsley. *Convergence of Probability Measures.* John Wiley and Sons, New York, 1968.

[4] Yu.V. Bulavsky, On choosing the survival factor in the Neumann-Ulam scheme. In: *Methods and Algorithms of Statistical Modelling*, Comp. Cent. Sib. Branch, USSR Acad. Sci., Novosibirsk, 1983, pp. 13–20 (in Russian).

[5] Yu.V. Bulavsky. Minimization of collision estimate variance in problems with alternating sign terms. *Zh. Vychisl. Mat. Mat. Fiz.* (1984) **1**, 148–152 (in Russian).

[6] N.N. Chentsov, Weak Convergence of Random Processes with Trajectories without Discontinuous of the Second Type, *Teor. Veroyatnost.i Primenen.* (1956) **1**, 154–161 (in Russian).

[7] A.F. Cheshkova. 'Walk on spheres' algorithms for solving Helmgoltz equation in the n-dimensional Euclidean space. *Bull. Novosibirsk Comp. Center, Num. Anal.* (1993) No. 4, pp. 7–18.

[8] W.A. Coleman, Mathematical verification of a certain Monte Carlo sampling technique and applications of the techniques to radiation transport problems. *Nucl. Sci. and Engng.* (1968) **32**, 76–81.

[9] J.H. Curtiss. Monte Carlo methods for the iteration of linear operators. *J. Math. Phys.* (1954) **32**, 209–232.

[10] N. Danford and J.T. Schwartz, *Linear Operators. Part I. General Theory.* Interscience Publishers, New York, 1958.

[11] B. Davison, *Neutron Transport Theory.* Oxford University Press, 1957.

[12] I.G. Dyad'kin and V.N. Starikov. On one possibility of machine-time saving in solving Laplace equation by the Monte Carlo method. *Zh. Vychisl. Mat. Mat. Fiz.* (1965) **5**, 936–938 (in Russian).

[13] B.S. Elepov, A.A. Kronberg, G.A. Mikhailov, and K.K. Sabelfeld, *Solution of Boundary-Value Problems by the Monte Carlo Method.* Nauka, Novosibirsk, 1980 (in Russian).

[14] B.S. Elepov and G.A. Mikhailov, On the solution of Dirichlet problem for equation $\Delta u + cu = -g$ by the 'walk on spheres' method. *Zh. Vychisl. Mat. Mat. Fiz.* (1969) **9**, No. 3, 647–654 (in Russian).

[15] B.S. Elepov and G.A. Mikhailov, Using fundamental solutions to elliptic equations for constructing Monte Carlo algorithms. *Zh. Vychisl. Mat. Mat. Fiz.* (1974) **14**, No. 3, 728–736 (in Russian).

[16] S.M. Ermakov. The Monte Carlo method for iterating nonlinear operators. *Dok. Akad. Nauk SSSR* (1972) **204**, No. 2, pp. 271–274 (in Russian).

[17] S.M. Ermakov and G.A. Mikhailov, *Statistical Modelling.* Nauka, Moscow, 1982 (in Russian).

[18] S.M. Ermakov, V.V. Nekrutkin and A.S. Sipin. *On the Use of Stochastic Processes for Solving Classical Equations of Mathematical Physics.* Nauka, Moscow, 1984 (in Russian).

[19] W. Feller, *An Introduction to Probability Theory and its Application, Vol.2.* Jonh Wiley and Sons, Inc., New York, 1971.

[20] A.S. Frolov and N.N. Chentsov, On Monte Carlo calculation of definite integrals dependent on a parameter. *Zh. Vychisl. Mat. Mat. Fiz.* (1962) **2**, 714–717 (in Russian).

[21] J.M. Hammersley and D.C. Handscomb, *Monte Carlo Methods.* Methuen, London, 1964.

[22] M.S. Ivanov, Numerical modelling of the dilute gas flows in transitional regime. *Ph. D Thesis,* Novosibirsk, 1979 (in Russian).

[23] H. Kahn, Use of different Monte Carlo sampling techniques. In: *Symposium on Monte Carlo Methods* (Ed. H.A. Meyer), Wiley, New York, 1956, pp. 146–190.

[24] M.H. Kalos and P.A. Whitlock, *Monte Carlo Methods. Vol. 1: Basics.* Jonh Wiley and Sons, New York, Chichester, Brisbane, Toronto, Singapore, 1986.

[25] R.Z. Khas'minskii, Probabilistic representation of solutions to some differential equations. In: *Theory of Probability and Mathematical Statistics,* Proceedings of the VIth All-Union Conference, Vilnyus, 1962 (in Russian).

[26] G.Z. Lotova. Calculation of time constant of particle breeding by Monte Carlo method using parametric derivatives. *Bull. Novosibirsk Comp. Center, Num. Anal.* (1993) No. 4, pp. 27–34.

[27] G.I. Marchuk and V.I. Agoshkov, Conjugated operators and perturbation algorithms in nonlinear problems. Perturbation algorithms. *Preprint No. 132,* Dept. Numer. Math., USSR Acad. Sci., Moscow, 1986 (in Russian).

[28] G.I. Marchuk, G.A. Mikhailov, M.A. Nazaraliev, *et al. The Monte Carlo Method in Atmospheric Optics.* Springer, Berlin–Heidelberg, 1980.

[29] G.A. Mikhailov, Monte Carlo calculations of critical systems. *Zh. Vychisl. Mat. Mat. Fiz.* (1966) **6**, 71–80 (in Russian).

[30] G.A. Mikhailov, On Calculations of Perturbations of the Nuclear Reactors by the Monte Carlo Method, *Zh. Vychisl. Mat. Mat. Fiz.* (1966) **6**, 380–384 (in Russian).

[31] G.A. Mikhailov, Monte Carlo estimation of some nonlinear functionals and approximate calculation of group constants in transport theory. *Zh. Vychisl. Mat. Mat. Fiz.* (1968) **8**, 590–599 (in Russian).

[32] G.A. Mikhailov, A method of simulating the free path of particle. *Atomn. Energiya* (1970) **28**, 175 (in Russian).

[33] G.A. Mikhailov, A new Monte Carlo algorithm for estimating the maximum eigenvalue of an integral operator. *Docl. Akad. Nauk SSSR* (1970) **191**, No. 5, 993–996 (in Russian).

[34] G.A. Mikhailov, Assessment of costs of modelling the 'walk on spheres' process for some domain types. *Zh. Vychisl. Mat. Mat. Fiz.* (1979) **19**, No. 2, 510–515 (in Russian).

[35] G.A. Mikhailov. Nonlinear equations, related to optimization of Monte Carlo methods for solving integral equations of the second kind. *Dokl. Akad. Nauk SSSR* (1980) **252**, No.4, pp. 792–796 (in Russian).

[36] G.A. Mikhailov, The variance of vector Monte Carlo algorithms. *Dokl. Akad. Nauk SSSR* (1980) **253**, No.5, 1047–1050 (in Russian).

[37] G.A. Mikhailov, Monte Carlo methods and perturbation theory. *Sov. J. Numer. Anal. Math. Modelling* (1988) **3**, No.1, 47–61.

[38] G.A. Mikhailov.Cost of Monte Carlo methods for global estimation of solutions to multidimensional problems. *Sov. J. Numer. Anal. Math. Modelling* (1991) **6**, No.2, 131–150.

[39] G.A. Mikhailov, *Minimization of Computational Costs of Nonanalogue Monte Carlo Methods.* World Scientific, Singapore, 1991.

[40] G.A. Mikhailov, *Optimization of Weighted Monte Carlo Methods.* Springer, Berlin–Heidelberg, 1992.

[41] G.A Mikhailov and S.G. Musikhin, Algorithms for Monte Carlo calculation of a complex reactor cell. *Zh. Vychisl. Mat. Mat. Fiz.* (1981) **10**, 432–440 (in Russian).

[42] G.A. Mikhalov, A.V. Voitishek. Convergence of numerical models of stochastic fields connected with Palm fluxes of points. *Dokl. Russian Akad. Nauk* (1994) **335**, No.3, pp. 291–294.

[43] M.E. Muller, Some continuous Monte Carlo methods for the Dirichlet problem. *Ann. Math. Stat.* (1956) **27**, No.3, 569–589.

[44] M.Yu. Plotnikov. Using the weighted Monte Carlo method for solving nonlinear integral equations. *Russ. J. Numer. Anal. Math. Modelling* (1994) **9**, No.2, pp. 1–24.

[45] S.I. Pokhozhayev. On the Dirichlet problem for the equation $\Delta u = u^2$. *Dokl. Akad. Nauk SSSR* (1960) **134**, No.4, pp. 769–773 (in Russian).

[46] R.T. Rockafellar. *Convex Analysis.* Princeton University Press, Princeton–New Jersey, 1970.

[47] K.K. Sabelfeld, *Monte Carlo Methods in Boundary-Value Problems.* Springer–Verlag, Berlin–Heidelberg–New York–London, 1991.

[48] A.F. Sizova. On the variance of collision estimate for solving nonlinear problems by Monte Carlo method. *Vestnik Leningradskogo Univ.* (1976), No.1, 152–155 (in Russian).

[49] V.V. Smelov, *Lectures on Transport Theory.* Atomizdat, Moscow, 1978 (in Russian).

[50] V.I. Smirnov, *A Course in Higher Mathematics.* Vol.4. Nauka, Moscow, 1974 (in Russian).

[51] J.N. Sneddon, *The Use of Integral Transforms.* McGrae-Hill, New York, 1972.

[52] J.M. Sobol, *Computational Monte Carlo Methods.* Nauka, Moscow, 1973 (in Russian).

[53] J. Spanier and E.M. Gelbard, *Monte Carlo Principles and Neutron Transport Problems.* Addison-Wesley, Reading, PA, 1969.

[54] G.A. Syrlybaeva, The proof of the finiteness of the variance of a Monte Carlo estimate for the solution of the Dirichlet problem. In: *Theory and Applications of Statistical Modelling,* Comp. Cent. Sib. Branch, USSR Acad. Sci., Novosibirsk, 1988, pp. 88–93 (in Russian).

[55] D.V. Tolstolytkin. A new Monte Carlo method for calculation of covariance function of the solution of biharmonic equation. *Bull. Novosibirsk Comp. Center, Num. Anal.* (1993) No.4, pp. 103–108.

[56] V.F. Turchin, V.Z. Nozik. Statistical regularization of solution of noncorrect problems. *Izv. Akad. Nauk SSSR, Ser. Fiz. Atmos. Okeana* (1969) **5**, 29–38.

[57] V.G. Zolotukhin and L.V. Maiorov, *Monte Carlo Assessment of Criticality Parameters of Reactors.* Energoatomizdat, Moscow, 1984 (in Russian).

Bei Fragen zur Produktsicherheit wenden Sie sich bitte an:
If you have any questions regarding product safety, please contact:

Walter de Gruyter GmbH
Genthiner Straße 13
10785 Berlin
productsafety@degruyterbrill.com